Advanced Structured Materials

Volume 48

Series editors

Andreas Öchsner, Southport Queensland, Australia
Lucas F. M. da Silva, Porto, Portugal
Holm Altenbach, Magdeburg, Germany

For further volumes:
http://www.springer.com/series/8611

J. M. P. Q. Delgado · A. G. Barbosa de Lima
Editors

Transport Phenomena and Drying of Solids and Particulate Materials

Springer

Editors
J. M. P. Q. Delgado
Laboratory of Building Physics
Department of Civil Engineering
Faculty of Engineering of University of Porto
Porto
Portugal

A. G. Barbosa de Lima
Department of Mechanical Engineering
Federal University of Campina Grande
Campina Grande
Brazil

ISSN 1869-8433 ISSN 1869-8441 (electronic)
ISBN 978-3-319-38147-3 ISBN 978-3-319-04054-7 (eBook)
DOI 10.1007/978-3-319-04054-7
Springer Cham Heidelberg New York Dordrecht London

Printed on acid-free paper

Springer is part of Springer Science+Business Media (www.springer.com)

Contents

Porous Materials Drying Model Based on the Thermodynamics of Irreversible Processes: Background and Application

A. G. Barbosa de Lima, J. M. P. Q. Delgado, V. A. B. de Oliveira, J. C. S. de Melo and C. Joaquina e Silva

Abstract This chapter focuses on the heat and mass transfer (drying) in capillary-porous bodies using both the mechanistic and non-equilibrium thermodynamic approaches. A new coupled mathematical modeling to predict heat and moisture (liquid and vapor) transfer in wet capillary-porous bodies with particular reference to prolate spheroidal solids is presented and discussed. The mathematical model is based on the thermodynamics of irreversible processes by considering variable transport coefficients and equilibrium or convective boundary conditions at the surface of the solid. All the partial differential equations presented in the model have been written in prolate spheroidal coordinates. The finite-volume method has been used to obtain the numerical solution of the governing equations. Application has been done to wheat kernel drying and comparison between predicted and experimental data has showed good agreement.

Keywords Drying · Numerical solution · Mass · Heat · Elliptical geometry

---

A. G. Barbosa de Lima (✉) · J. C. S. de Melo · C. Joaquina e Silva
Department of Mechanical Engineering, Federal University of Campina Grande,
Av. Aprígio Veloso, 882 Bodocongó, Campina Grande, PB Zip Code: 58429-900, Brazil
e-mail: gilson@dem.ufcg.edu.br

J. C. S. de Melo
e-mail: jcarlosmequi@yahoo.com.br

C. Joaquina e Silva
e-mail: carlota.jsilva@gmail.com

J. M. P. Q. Delgado
LFC – Building Physics Laboratory, Civil Engineering Department Faculty of Engineering,
University of Porto, Portò, Portugal
e-mail: jdelgado@fe.up.pt

V. A. B. de Oliveira
Department of Civil Engineering, State University of Paraíba, Araruna, PB, Brazil
e-mail: vitaloliveira@uepb.edu.br

J. M. P. Q. Delgado and A. G. Barbosa de Lima (eds.), *Transport Phenomena and Drying of Solids and Particulate Materials*, Advanced Structured Materials 48,
DOI: 10.1007/978-3-319-04054-7_1,

1 Introduction

The diffusion phenomenon occurs in many industrial processes, such as drying, wetting, heating and cooling. These processes have been applied to preservation and conservation of different biological products, mainly foodstuffs (grains, fruits, vegetables, etc.) well known as hygroscopic capillary-porous bodies.

Drying is one of the most important processes used in the processing of foods and in the storage of grains. This process consists of the partial transfer of the liquid part (usually water) of a wet solid. The drying process can also be explained as a process of heat and mass transfer that generates the removal through evaporation of part of the moisture contained in the product [1] and takes place simultaneously with geometrical variations (shrinkage, dilation and deformation) and physical and micro-structural modifications (color, flavour, appearance, nutrients, germination, depending on the product species). Drying differs from other separation techniques due to the movement of the molecules, which in this case is obtained by a mass transfer of the liquid and/or vapor due to the difference in partial pressure of the steam between the surface of the solid to be evaporated and the air that surrounds it. In the case of foods, water must be removed from the moist material up to a level where deterioration provoked by microorganisms can be minimised, in order to maintain quality characteristics these products that determine its acceptance and commercialization. However, drying rate is affected by many factors such as velocity, temperature and relative humidity of air, type of solid and dryer, drying method and initial moisture content of solid.

For accurately describe moisture migration within biological products, mainly foodstuffs (grains, fruits, vegetables, etc.) and to explain the effects of drying on the quality of the material, the heat and moisture transfer within a individual particles must be understood and accurately represented by a mathematical model. From a review of the literature, it is noted that several researches prefer the diffusion model applied only to known geometric shapes specifically: plates, cylinders, spheres and parallelepipeds [2–5], and for prolate and oblate spheroidal bodies [6–14] with specified boundary conditions and constant thermo-physical properties assuming moisture migration inside the solid by liquid or vapor diffusion only. However, other researchers shown that the moisture migrates in both liquid and vapor phases can occur within the solid, due to concentration gradients, partial vapor-pressure gradients, capillary forces, differences in total pressure and gravity [1, 15–27].

According to Luikov [19], transfer of vapor and inert gas can take place by molecular means in the form of diffusion and by molar means due to filtration motion of the whole steam-gas mixture within the porous body under the influence of a fall in aggregate pressure. Therefore, the derivation of a model of mass transfer in capillary-porous bodies on the basis of molecular and molar transfer mechanism presents great difficulties. In this sense, Fortes [1], Fortes and Okos [24], Fortes et al. [28] present a set of transport equations that incorporates most of the models presented in the literature by combining the mechanistic and irreversible thermodynamics approaches to describe heat and mass transfer in porous media. In the model is postulated to be valid the Gibb's equation, linear phenomenological laws, Onsager's fundamental theorem, Curie's principle and local thermodynamic equilibrium. Based on the thermodynamic concept only, the authors has proposed that the driving-force for moisture migration in the liquid and vapor phases is the local equilibrium moisture content gradient instead of the moisture content as reported in the literature (Fick's second law of diffusion). Thus, water may migrates from region of lower to higher moisture content, since the region of higher moisture content is at a lower equilibrium moisture content and the temperature gradient term is greater than the equilibrium moisture content gradient term and favors this movement. Under isothermal condition, water always moves from regions of higher to lower equilibrium moisture content. As related by Fortes and Okos [24] all the physical parameters cited in the model can be obtained from experimental techniques.

How complement for these researches, in this chapter, a two-dimensional numerical analysis is performed for simultaneous heat and mass (liquid and vapor) transport in bodies with ellipsoidal shape, using the Fortes and Oko's theory (non-equilibrium thermodynamic approaches).

2 The Macroscopic Governing Equation

2.1 General Distributed Model

In the model which will be presented here, the following considerations were made:

(a) The solid is considered to be homogeneous and continuum;
(b) The shape of solid is approached to be an ellipsoid of revolution;
(c) The body is axially symmetric around z-axis;
(d) The solid is composed of water in liquid and vapor phases and solid matter;
(e) The shrinkage and dilation are neglected;
(f) The unique mechanism of drying is diffusion;
(g) Convection phenomenon within the solid is negligible.
(h) Moisture migration (liquid and vapor) due to gravity is negligible;
(i) The liquid and vapor diffusion phenomenon occurs under falling drying rate;
(j) Thermo-physical properties are variables during drying process;
(k) The moisture content and temperature fields are considered symmetric around z-axis at any time and constant in the beginning of the process;
(l) The phenomenon occurs under equilibrium or convective boundary condition at the surface;
(m) No energy and mass generation occurs within the solid.

2.1.1 Mass Diffusion Model

The moisture transport within the solid was assumed to be in the liquid and vapor states and governed by the transient liquid and vapor diffusion equation. For the model of mass transfer inside the wet porous solid we have the following equations [1, 24, 27, 28]:

- Liquid flux:

$$\vec{J}_\ell = -\rho_\ell k_\ell R_v \ell n H \nabla T - \rho_\ell k_\ell \frac{R_v T}{H} \frac{\partial H}{\partial M} \nabla M + \rho_\ell k_\ell \vec{g} \tag{1}$$

- Vapor flux:

$$\vec{J}_v = -k_v \left(\rho_{vo} \frac{\partial H}{\partial T} + H \frac{d\rho_{vo}}{dT} \right) \nabla T - k_v \rho_{vo} \frac{\partial H}{\partial M} \nabla M + \left(\frac{H k_v \rho_{vo}}{T R_v} \right) \vec{g} \tag{2}$$

where M (dry basis) should be taken as the equilibrium moisture content in any location inside the porous media. In this equations $\vec{g}$ is the gravitational acceleration vector; H is the relative humidity; k_l is the liquid conductivity; k_v is the vapor conductivity; R_V is the universal gas constant as applied to air; T is the air temperature; q is the density.

Assuming that no ice is present, that the mass of air is neglected, and that mass of vapor is neglected in comparison with the mass of liquid (but not its flux) [19], one mass balance in an differential control volume leads to:

$$\frac{\partial(\rho_s M)}{\partial t} = -\nabla \cdot \left(\vec{J}_\ell + \vec{J}_v \right). \tag{3}$$

2.1.2 Heat Conduction Model

For the model of heat transfer inside the wet porous solid, we have the following energy equation [1, 24, 27, 28]:

$$\frac{\partial}{\partial t}(\rho_s c_b T) - \frac{\partial}{\partial t}(\rho_s L_w M) = -\nabla \cdot \vec{J}_q - \nabla \cdot \left(L_v \vec{J}_v \right) - \vec{J}_\ell \cdot c_\ell \nabla T - \vec{J}_v \cdot c_v \nabla T \tag{4}$$

where the c is the specific heat; L_w is the specific differential heat of wetting and L_v is the specific latent heat of vaporization. The heat flux is given by:

$$\begin{aligned} \vec{J}_q = & - k_T \nabla T - \left[\rho_\ell k_\ell R_v \ell n H + k_v \left(\rho_{vo} \frac{\partial H}{\partial T} + H \frac{d\rho_{vo}}{dT} \right) \right] \frac{R_v T^2}{H} \frac{\partial H}{\partial M} \nabla M \\ & + T \left[\rho_\ell k_\ell R_v \ell n H + k_v \left(\rho_{vo} \frac{\partial H}{\partial T} + H \frac{d\rho_{vo}}{dT} \right) \right] \vec{g} \end{aligned} \tag{5}$$

where k_T is the thermal conductivity. More details about the notation may be found in the references cited in the text.

2.2 Conservation Equations Applied to Prolate Spheroidal Solids

For better understanding of the problem treated herein, we must consider the diffusion phenomenon in an ellipsoid of revolution, in which the axis of revolution is greater than the other axis as pictured in Fig. 1. In this case it is said to be prolate spheroid, while, if the axis of revolution is smaller than the other, it is called oblate spheroid. Further, if the axis of revolution is equal to the other axis, the solid is called sphere.

Based in the shape of the solid to be studied (Fig. 1), is better to use a new orthogonal coordinate system ξ, η, ζ instead of the Cartesian coordinate x, y and z. In the case of the body with ellipsoidal geometric shape, an adequate coordinate system is the prolate spheroidal coordinate system (Fig. 2). The mathematical relations between Cartesian and prolate spheroidal coordinate systems are given by [29–31]:

$$x = L\sqrt{(1-\xi^2)(\eta^2-1)}\zeta \quad (6)$$

$$y = L\sqrt{(1-\xi^2)(\eta^2-1)}\sqrt{(1-\zeta^2)} \quad (7)$$

$$z = L\,\xi\,\eta \quad (8)$$

where $\xi = \cosh\mu$; $\eta = \cos\phi$; $\zeta = \cos\omega$ and $L = (L_2^2 - L_1^2)^{1/2}$ is the focal length.

According to Fig. 2, the surfaces $\xi = \xi_o$ (constant), $\xi_o > 1$ is an elongated ellipsoid of revolution with major axis of length $L\xi$ and minor axis of length $L(\xi^2 - 1)^{1/2}$. The surfaces ξ constant are a confocal family of prolate spheroids having their common center at the origin. The degenerate surface $\xi = 1$ is the curve that link the center ($z = 0$) to the focal point ($z = L$). The surface $\eta = \eta_o$ (constant), $\eta_o < 1$, is a hyperboloid of revolution of two sheets with an asymptotic cone whose generating by line passes through the origin and is inclined at the angle $\phi = \cos^{-1}\eta$ to the z-axis. The degenerate surface $\eta = 1$ is part of the axis $z > L$.

Then, calculating the metric coefficients of the transformation of coordinates, the volume of the body, the gradient and the Laplacian for the new coordinate system, utilizing the symmetry around the z-axis, $\partial/\partial\omega = 0 \Rightarrow \partial/\partial\zeta = 0$ and using the considerations presented, the governing mathematical equations can be formulated in the prolate spheroidal coordinate system [6, 8, 9, 15–17, 29].

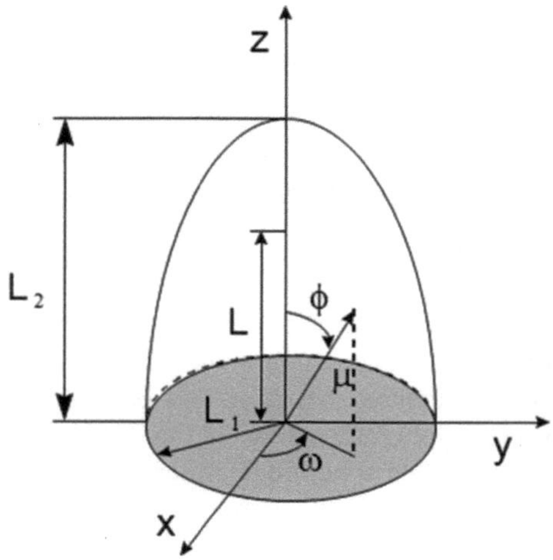

Fig. 1 Geometrical parameters and shape of a prolate spheroidal solid

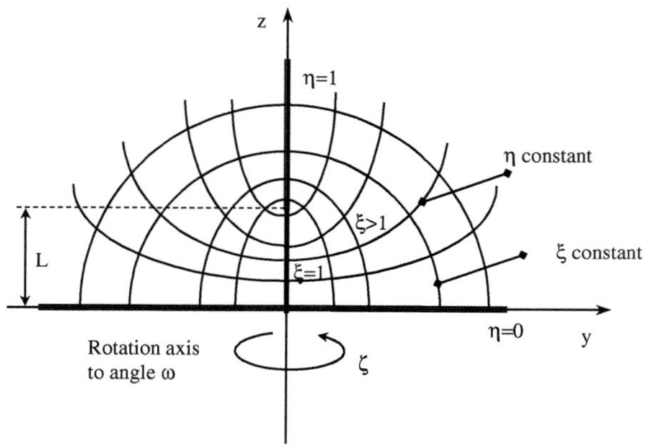

Fig. 2 Prolate spheroidal coordinate system

The mass transfer equation, Eq. (3), written in prolate spheroidal coordinates is:

$$\begin{aligned}\frac{\partial(\rho_s M)}{\partial t} &= \frac{1}{L^2(\xi^2-\eta^2)}\left\{\frac{\partial}{\partial \xi}\left[(\xi^2-1)\Gamma_1^{\Phi}\frac{\partial M}{\partial \xi}\right]+\frac{\partial}{\partial \eta}\left[(1-\eta^2)\Gamma_1^{\Phi}\frac{\partial M}{\partial \eta}\right]\right\}\\ &+\frac{1}{L^2(\xi^2-\eta^2)}\left\{\frac{\partial}{\partial \xi}\left[(\xi^2-1)\Gamma_2^{\Phi}\frac{\partial T}{\partial \xi}\right]+\frac{\partial}{\partial \eta}\left[(1-\eta^2)\Gamma_2^{\Phi}\frac{\partial T}{\partial \eta}\right]\right\}\end{aligned} \tag{9}$$

where

$$\Gamma_1^{\varphi} = \left(\rho_{\ell} k_{\ell} \frac{R_v T}{H} \left(\frac{\partial H}{\partial M} \right) + k_v \rho_{v0} \left(\frac{\partial H}{\partial M} \right) \right); \tag{10}$$

$$\Gamma_2^{\varphi} = \left(\rho_{\ell} k_{\ell} R_v \ln(H) + k_v \left(\rho_{v0} \frac{\partial H}{\partial T} + H \frac{d\rho_{v0}}{dT} \right) \right); \tag{11}$$

To complete the mathematical formulation, the following initial and boundary conditions are used:

- Free surface: the mass diffusive flux is equal to the mass convective flux at the surface of the solid.

$$\left(\vec{J}_{\ell} + \vec{J}_v \right) \Big|_{\xi = \xi_f} = h_m \left(M_{\xi = \xi_f} - M_e \right) \tag{12}$$

where $n_f = L_2/L$ at the surface

- Planes of symmetry: the angular and radial gradients of moisture content are equals to zero at the planes of symmetry.

$$\frac{\partial M(\xi; \quad \eta = 1; \quad t)}{\partial \eta} = 0 \tag{13}$$

$$\frac{\partial M(\xi; \quad \eta = 0; \quad t)}{\partial \eta} = 0 \tag{14}$$

$$\frac{\partial M(\xi = 1; \quad \eta; \quad t)}{\partial \xi} = 0 \tag{15}$$

- Initial condition in the interior of the solid

$$M(\xi; \quad \eta; \quad t = 0) = M_o \tag{16}$$

The average moisture content of the body at any instant was calculated as follows:

$$\overline{M} = \frac{1}{V} \int_V M dV \tag{17}$$

where V is the volume of the ellipsoidal body calculated as follows:

$$V = \frac{4}{3} \pi L_2 L_1^2 \tag{18}$$

The energy equation, Eq. (4), written in prolate spheroidal coordinates is:

$$\frac{\partial}{\partial t}(\rho_s c_b T)-\frac{\partial}{\partial t}(\rho_s L_w M)=\frac{1}{L^2(\xi^2-\eta^2)}\left\{\frac{\partial}{\partial\xi}\left[(\xi^2-1)\Gamma_3^\Phi\frac{\partial T}{\partial\xi}\right]+\frac{\partial}{\partial\eta}\left[(1-\eta^2)\Gamma_3^\Phi\frac{\partial T}{\partial\eta}\right]\right\}$$
$$+\frac{1}{L^2(\xi^2-\eta^2)}\left\{\frac{\partial}{\partial\xi}\left[(\xi^2-1)\Gamma_5^\Phi\frac{\partial T}{\partial\xi}\right]+\frac{\partial}{\partial\eta}\left[(1-\eta^2)\Gamma_5^\Phi\frac{\partial T}{\partial\eta}\right]\right.$$
$$\left.+\frac{\partial}{\partial\xi}\left[(\xi^2-1)\Gamma_4^\Phi\frac{\partial M}{\partial\xi}\right]\right\}+\frac{1}{L^2(\xi^2-\eta^2)}$$
$$\left\{+\frac{\partial}{\partial\eta}\left[(1-\eta^2)\Gamma_4^\Phi\frac{\partial M}{\partial\eta}\right]+\frac{\partial}{\partial\xi}\left[(\xi^2-1)\Gamma_6^\Phi\frac{\partial M}{\partial\xi}\right]+\frac{\partial}{\partial\eta}\left[(1-\eta^2)\Gamma_6^\Phi\frac{\partial M}{\partial\eta}\right]\right\} \tag{19}$$
$$+\frac{\Gamma_7^\Phi}{L^2(\xi^2-\eta^2)}\left[(\xi^2-1)\left(\frac{\partial T}{\partial\xi}\right)^2+(1-\eta^2)\left(\frac{\partial T}{\partial\eta}\right)^2\right]$$
$$+\frac{\Gamma_8^\Phi}{L^2(\xi^2-\eta^2)}\left[(\xi^2-1)^2\frac{\partial T}{\partial\xi}\frac{\partial M}{\partial\xi}+(1-\eta^2)\frac{\partial T}{\partial\eta}\frac{\partial M}{\partial\eta}\right]$$

where

$$\Gamma_3^\Phi = k_T; \tag{20}$$

$$\Gamma_4^\Phi = \left[\rho_\ell k_\ell R_v \ell n(H) + k_v\left(\rho_{v0}\frac{\partial H}{\partial T}+H\frac{d\rho_{v0}}{dT}\right)\right]\left(\frac{R_v T^2}{H}\frac{\partial H}{\partial M}\right); \tag{21}$$

$$\Gamma_5^\Phi = L_v k_v\left(\rho_{v0}\frac{\partial H}{\partial T}+H\frac{d\rho_{v0}}{dT}\right); \tag{22}$$

$$\Gamma_6^\Phi = L_v k_v \rho_{v0}\left(\frac{\partial H}{\partial M}\right); \tag{23}$$

$$\Gamma_7^\Phi = c_\ell\rho_\ell k_\ell R_v \ell n(H) + k_v c_v\left(\rho_{v0}\frac{\partial H}{\partial T}+H\left(\frac{d\rho_{v0}}{dT}\right)\right); \tag{24}$$

$$\Gamma_8^\Phi = c_\ell\rho_\ell k_\ell\frac{R_v}{H}\left(\frac{\partial H}{\partial M}\right)+k_v c_v\rho_{v0}\frac{\partial H}{\partial M}. \tag{25}$$

The initial and boundary conditions are:

- Free surface: the heat convective flux supplied to the body surface equals to the heat diffusive flux plus the energy necessary to evaporate the liquid water and to heat the vapor produced at the surface of the prolate spheroid from surface temperature to the drying-air temperature.

$$\vec{J}_q\Big|_{\xi=\xi_f} = h_c\left(T_a - T_{\xi=\xi_f}\right)+L_v\vec{J}_\ell+\left(\vec{J}_\ell+\vec{J}_v\right)c_v\left(T_a-T_{\xi=\xi_f}\right) \tag{26}$$

- Planes of symmetry: the angular and radial gradients of temperature are equals to zero at the planes of symmetry.

$$\frac{\partial T(\xi;\quad \eta = 1;\quad t)}{\partial \eta} = 0 \tag{27}$$

$$\frac{\partial T(\xi;\quad \eta = 0;\quad t)}{\partial \eta} = 0 \tag{28}$$

$$\frac{\partial T(\xi = 1;\quad \eta;\quad t)}{\partial \xi} = 0 \tag{29}$$

- Initial condition inside the solid

$$T(\xi;\quad \eta;\quad t = 0) = T_0 = cte \tag{30}$$

The average temperature of the body during diffusion phenomenon was calculated as follows:

$$\bar{T} = \frac{1}{V}\int_V T dV \tag{31}$$

This heat and mass transfer model is interesting and complex because it incorporates the mechanistic [32–35] and the irreversible thermodynamic approaches [18, 19, 21, 36, 37]. More details about the heat and mass transfer model reported here may be found in Fortes [1], Fortes et al. [28] and Oliveira [38].

3 Computational Strategy for Solution of the Governing Equations and Simulation

Various numerical methods have been used to solve the problem of transient diffusion, such as, finite-difference, finite-element, boundary-element and finite-volume methods. Several discussions may be encountered in previous reports [39–42]. In particular, in this work, we used the finite-volume method. In the simulation of diffusion phenomenon in prolate spheroids was utilized a certain domain, due to the symmetry of the body. Assuming fully implicit formulation, the Eqs. (9) and (17) were integrated in the volume and time. How results the following discretization equation was obtained:

$$A_P\Phi_P = A_E\Phi_E + A_W\Phi_W + A_N\Phi_N + A_S\Phi_S + A_P^o\Phi_P^o + B \tag{32}$$

where Φ represents M or T and B is the source term. Details about the discretization procedure can be found in the references cited in the text.

Equation (32) constitutes a set of linear algebraic equation which must be solved, in order for obtain the kinetic values and distribution of the potential Φ inside the solid at any instant. The method employed to solve it was the Gauss-Siedel iterative method. Convergence was assumed to have been reached when the following absolute value is obtained in all nodal points:

$$\left|\Phi^{k+1} - \Phi^{k}\right| \leq 10^{-8} \tag{33}$$

where Φ represents M or T and k represent kth iteration.

In the discretization, the interface properties and non linearity (source-term linearization) were treated according to Patankar [40] and Maliska [41]. A computational code utilizing the Mathematica® commercial software was written to solve the set of equations for the moisture content and temperature profile and to determine the average moisture content and temperature.

The application of the finite-volume numerical method is seriously affected by the Δt values and the number of grid points utilized in the numerical calculations. In order to verify grid size and time step independence, results were obtained with four grid sizes and time steps. After this study, it was chosen a mesh with 20×20 nodal points and $\Delta t = 1$ s to be used in the iterative scheme. Figure 3 shows the numerical grid generated.

According to Lima and Nebra [6], for prolate spheroids, the shape of the mesh varies with L_2/L_1. When $L_2/L_1 \rightarrow \infty$, the focal point is dislocated to the surface of the body; the inverse occurs for $L_2/L_1 \rightarrow 1$, when the focal point tends to coincide with the geometrical center of the body. On the zy plane (physical plane), the control volumes are not uniformly spaced, presenting a higher concentration near the surface of the body relative to ξ (radial coordinate), and on the y axis relative to η (angular coordinate). Then, an irregular grid is obtained; however, in the $\xi\eta$ plane (computational plane), we have a regular grid. This regular grid has the property that the ratio of lengths of any two adjacent intervals is a constant and equal to one. Just in this plane all the numerical calculations are performed.

In the numerical stability analysis performed, only the finite-volume approach inside the solid was considered, and we does not have mechanisms to predict the influence of the boundary conditions on the numerical stability. The boundary values have large influence on grid spacing only at the nodal points immediately near the surface of the body, in spite of the elliptic character of Eqs. (9) and (19).

4 Application

4.1 Physical Problem Description: Wheat Grain Drying

Several biological products has shape approximately ellipsoidal, in the particular case, prolate spheroidal. As examples, we can cite rice, wheat, orange, silkworm cocoon and banana. The problem of heat and mass transfer in a prolate spheroid it

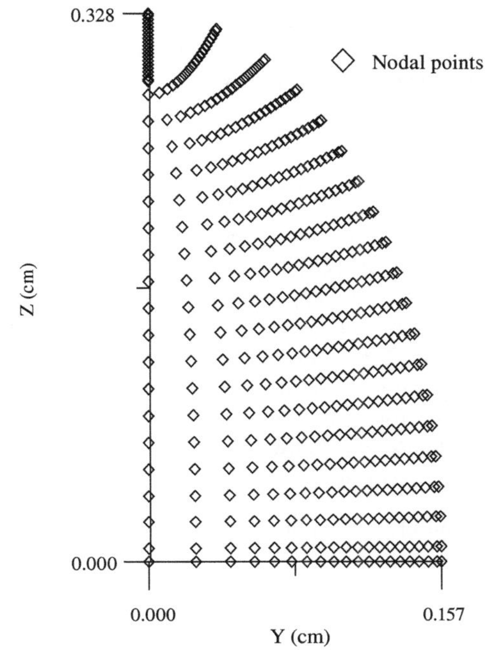

Fig. 3 Numerical grid in the physical plane

is of great interest because, it represents a high degree of precision of the problem in comparison to the approach this problem for cases of sphere or cylinder as used today. Since that sphere and cylinder are particular cases of prolate spheroids, it is now unnecessary to repeat the complex derivation for each different geometric shape.

Grain drying and storage are of great importance to the agricultural industry. Each year large quantities of grains are dried and placed in storage. The removal of moisture from agricultural crops has been subject of considerable research.

Wheat is a cereal crop produced worldwide; today it is grown all over the world, with different varieties shown according to the various climates.

Globally, wheat is an important human energy source and is one of the major staple foods because of its agronomical adaptability, ability of its flour to be made into various food material and ease of storage [43]. Wheat grain must be dried to 13.5 percent for immediate sale since over drying reduces the total weight of grain to be sold. However, if wheat is to be stored for any period of time, it is dried to 12 to 12.5 percent moisture content to prevent spoilage.

Wheat production is a direct function of high yielding varieties, chemical, fertilizers, mechanization and other energy inputs [44]. In Brazil, the average annual production of wheat in 2011 comes to about 5.69 million tones, against World's average of 701.40 million tones in the same year. Thus, Brazil presents a reduced contribution to the World wheat production. In the year 2011, the area under wheat cultivation in Brazil, comes to about 2.17 million ha which is many reduced in comparison with the World's wheat area [45].

Now, based on the information cited before, as an application we use the modeling reported here to predict drying of wheat grain using two boundary conditions: equilibrium (model 1) and convection (model 2) at the surface of the solid.

4.2 Experimental Data

Fortes et al. [28] report an experimental drying study of soft red winter wheat (Arthur). The apparatus used for drying consisted of a climate chamber with electric heaters, which allowed the temperature to be adjusted and controlled up to 150 °C. From this research were obtained average moisture content and centre temperature data of wheat kernel. Other details about the experimental procedure can be found in [28].

Herein, the following thermo-physical properties and geometrical parameters of wheat grain were used in the drying simulation:

- Saturated vapor density [28]

$$\rho_{vo} = (2.54 \times 10^8/T)\exp(-5200/\mathrm{T})\ \left(\mathrm{kg/m^3}\right) \tag{34}$$

  valid in $283.15 \leq T \leq 363.15\,\mathrm{K}$
- Water density [28]:

$$\rho_\ell = 1000\,\mathrm{kg/m^3}$$

- Dry solid density [28]:

$$\rho_s = 1265\,\mathrm{kg/m^3}$$

- Heat of vaporization of water in the wheat kernel [28]:

$$L_o = 3.11 \times 10^6 - 2.38 \times 10^3 T\ (\mathrm{J/kg}) \tag{35}$$

$$L_w = R_v T^2/H \times \partial H/\partial T \quad (\mathrm{J/kg}) \tag{36}$$

$$L_v = L_w + L_o \quad (\mathrm{J/kg}) \tag{37}$$

  valid in $273.15 \leq T \leq 363.15\,\mathrm{K}$. In the Eqs. (35)–(37) L_o represent the heat of vaporization of free water and L_w is the differential heat of adsorption.
- Isotherm equation (modified Henderson's equation) [28]:

$$H = 1 - \exp\left(-5869T^{-0.7750}M^{5203}T^{-1.363}\right)\ (\mathrm{decimal}) \tag{38}$$

  valid in $\begin{cases} 0 \leq M \leq 0.25(d.b.) \\ 298.15 \leq T \leq 363.15\,\mathrm{K} \end{cases}$

- Thermal conductivity [46]:

$$k_T = 0.1170 + 0.0011319 \times (100M)/(1+M)\ (\text{W/m K}) \tag{39}$$

valid in $0 \leq M \leq 0.25$ $(d.b.)$

- Liquid and vapor conductivities [28]:

$$k_\ell = a_1 \times 4.366 \times 10^{-18} H^3 \exp(-1331/T)\ (\text{s}) \tag{40}$$

$$k_v = a_2 \times 6.982 \times 10^{-9} (T - 273.16)^{0.41} \left(H^{0.1715} - H^{1.1715}\right) \left(\text{m}^2/\text{s}\right) \tag{41}$$

where a_1 and a_2 are constant to be determined by comparison of predicted results with experimental data. From Eqs. (40) and (41) both k_ℓ and k_v are dependent on relative humidity and temperature of air.

- Specific heat of moist wheat [47], and water in the liquid and vapor phases [46].

$$c_b = [1.394 + 0.0409(100M)/(1+M)] \times 10^{+3}\ (\text{J}/(\text{kg K}) \tag{42}$$

valid in $0 \leq M \leq 0.25$ $(d.b.)$.

$$c_\ell = 4218.69\ \text{J/kg K}$$
$$c_v = 1919.85\ \text{J/kg K}$$

- Dimensions of wheat kernel [46]:

$$L_1 = 1.5748\ \text{mm}$$
$$L_2 = 3.2760\ \text{mm}$$

By considering the grain like sphere with volume equivalent to the ellipsoid, convective heat transfer coefficient was obtained as follows [28]:

$$h_c = \frac{k_a}{d_p}\left(2 + 0.6 R_e^{1/2} P_r^{1/3}\right) \tag{43}$$

where k_a is the thermal conductivity of the air, d_p represents the equivalent diameter, and Re and Pr are Reynolds and Prandtl numbers, respectively.

Table 1 shows the air and material conditions used in this work.

The liquid and vapor conductivities (parameters a_1 and a_2) and the convective mass transfer coefficient were estimated using the least square error technique, as follows:

Table 1 Air and wheat experimental conditions used in this work [28]

Air			Wheat					t
T (°C)	H (%)	v (m/s)	M_o (d.b.)	M_{final} (d.b.)	M_e (d.b.)	T_o (°C)	T_e (°C)	(min)
87.8	5.6	1.71	0.2110	0.127	0.0165	26.0	87.8	1020

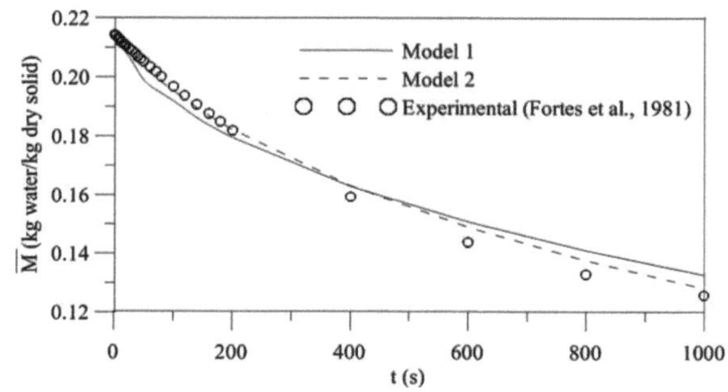

Fig. 4 Comparison between predicted and experimental dimensionless average moisture content as a function of drying time

$$ERMQ = \sum_{i=1}^{n} \left(\bar{M}_{i,Num} - \bar{M}_{i,Exp}\right)^2 \tag{44}$$

$$\bar{S}^2 = \frac{ERMQ}{(n - \hat{n})} \tag{45}$$

where ERMQ is the least square error, $\bar{S}^2$ is the variance, n is the number of experimental points and $\hat{n}$ is the fitted parameters number [48]. The starting point was the mathematical expression reported by Fortes et al. [28], to k_l and k_v applied to wheat by considering it like sphere. These equations were corrected to prolate spheroidal solid multiplying it by a constant parameter to be obtained by fitting.

4.3 Numerical Results and Analysis

4.3.1 Drying and Heating Kinetics

Figure 4 illustrates predicted results using the cited models compared with experimental data of the average moisture content obtained during the wheat drying reported by Fortes et al. [28]. From the analysis of this figure it can be see that a good agreement was obtained in two used models. Thus, both models accurately predict well the drying data. However, better results were obtained when more appropriate convective boundary condition was used (model 2). Figure 5 shows the average temperature of the wheat during drying process. It is verified that wheat almost reach the equilibrium temperature in ≈ 50 s for model 1

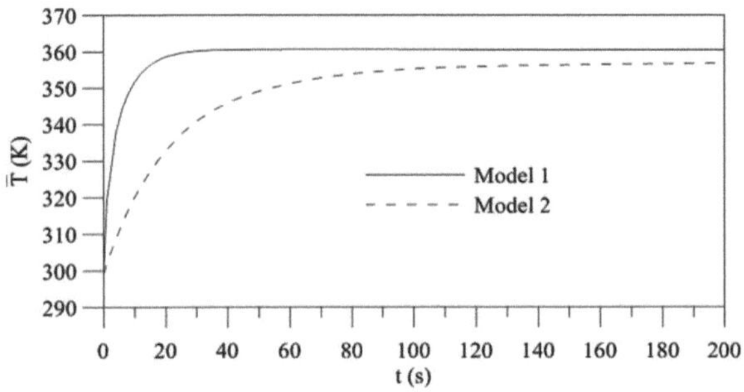

Fig. 5 Predicted average temperature of the wheat during the drying at 87.8 °C

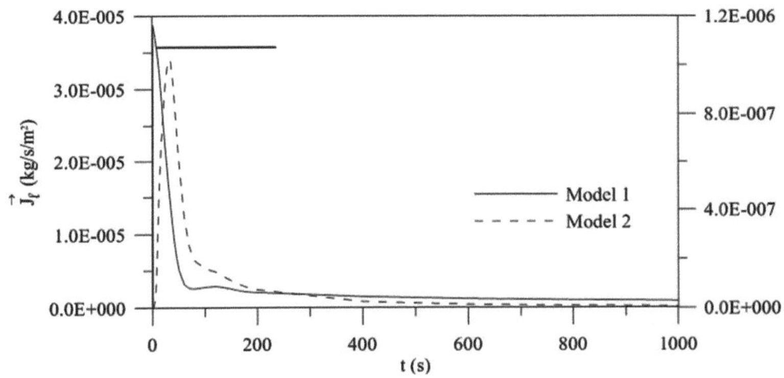

Fig. 6 Predicted liquid flux at the wheat kernel surface during the drying at 87.8 °C

Fig. 7 Predicted vapor flux at the wheat kernel surface during the drying at 87.8 °C

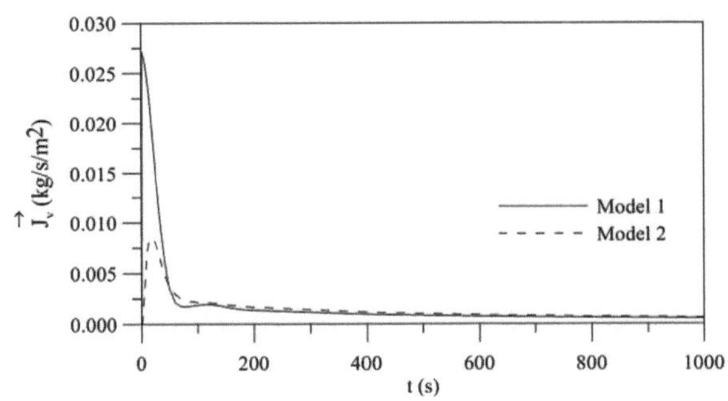

and $\approx$200 s for model 2, due to the boundary conditions used in this work; these results are in agreement with experimental data [28]. At these instant the wheat grain still very moist. The drying practically is starting. This period is so called as accommodation period of drying.

Figures 6, 7, 8 and 9 illustrate the behavior of the liquid and vapor fluxes at the surface of the solid along the drying. The numerical results showed that, vapor flux

Fig. 8 Predicted vapor and liquid fluxes relationships at the wheat kernel surface during the drying at 87.8 °C

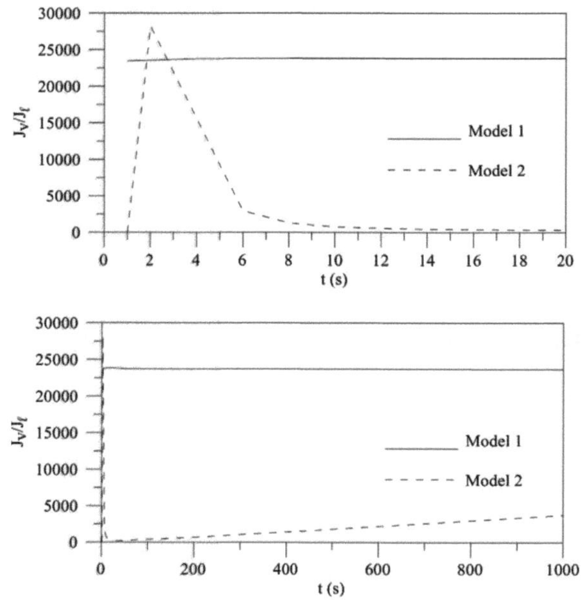

Fig. 9 Predicted total mass flow rate at the wheat kernel surface during the drying at 87.8 °C

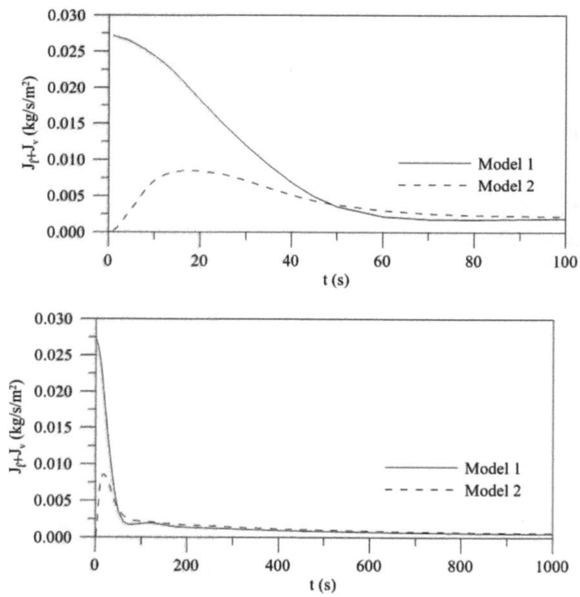

is dominant at the surface of the solid during the whole drying process in the air conditions used in this work, in both used models, being higher when more severe boundary conditions is used (equilibrium boundary condition, model 1). This behavior is resulted of the low initial moisture content of the solid and high drying-air temperature and lower air relative humidity. When the solid has high initial moisture content, for example, fruits and vegetables both vapor and liquid migrations are important within the solid. Thus, an assumption of either liquid or vapor flux, as the dominant mechanism for moisture migration in wet solid is unrealistic physically. These comments are in accordance with the literature. Hartley [49] and Fortes [1] report that the theory suggest and experiments evidence that such vapor movement can explain many trends observed experimentally in the drying of moist soils.

From the analysis of these figures and results for heat flux we states that heat, liquid and vapor fluxes due to temperature gradients are negligible with respect to those due to equilibrium moisture content gradients for elapsed time more than 200 s. Further, heat flux used to evaporate moisture at the surface of the grain must not be negligible with respect to the conductive heat flux in the short elapsed time (first stages of drying).

4.3.2 Moisture Content and Temperature Distributions

The moisture content distribution inside the solid is very important in order to study the evolution of hydric stress developed in the body due to the high moisture gradients. Figures 10 and 11 show the moisture content distribution inside the wheat grain exposed to the drying for 600 and 1000 s, for the models 1 and 2, respectively. From the analysis of these figures we notice a very similar drying behavior. It can be seen that the highest moisture content gradients happen near the surface. There is a strong drying also closed to the focal point. Then, these areas are more susceptible to have troubles such as cracks and fissures, due to higher moisture gradients and shrinkage velocity. It can also be seen that a moving evaporation front occurs from the surface to center of the solid.

Figures 12 and 13 present the temperature distribution inside of the wheat kernel after having heated up for 10 and 20 s, for the models 1 and 2, respectively. The highest temperature gradients and dilation velocity happen in the proximity of the focal point, turning this area more susceptible to thermal stresses, which will contribute to the appearance of deformations, cracks and fissures, besides supplying to the product the possibility to alter its color, texture and flavor, possibly reducing its quality. In every drying time, may be observed that the behavior of the isothermal lines is similar from the presented by the constant moisture content lines inside the solid. Being compared the temperature gradients with the moisture content gradients inside the wheat grain (Figs. 10, 11), it can be verified that these last ones are more pronounced, and the drying occurs practically under isothermal condition for $t > 50$ s of elapsed time. Besides, all liquid water eventually turns into vapor inside the grain or at the surface, thus varying both the local liquid and vapor fluxes within the wheat kernel. Further, we states that surface evaporation velocity is limited by the heat received from the surrounding.

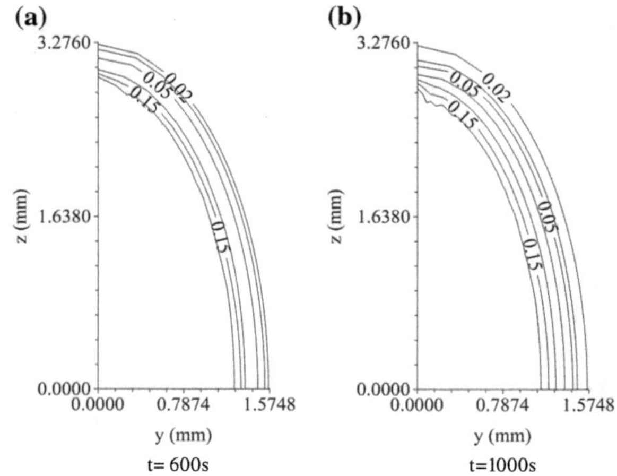

Fig. 10 Predicted moisture content distribution (dry basis) inside the wheat kernel during the drying at 87.8 °C (Model 1). **a** t = 600 s. **b** t = 1000 s

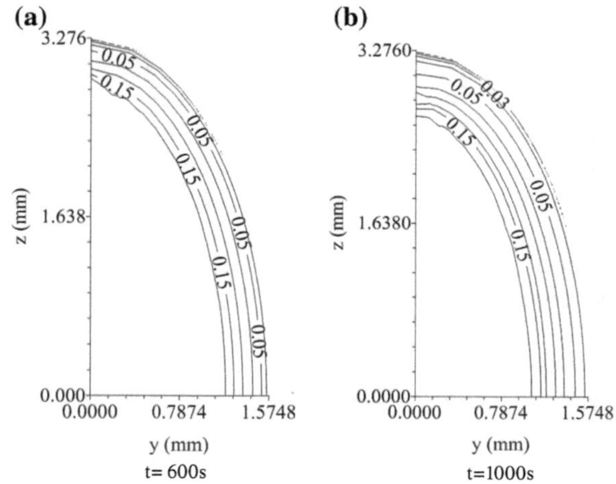

Fig. 11 Predicted moisture content distribution (dry basis) inside the wheat kernel during the drying at 87.8 °C (Model 2). **a** t = 600 s. **b** t = 1000 s

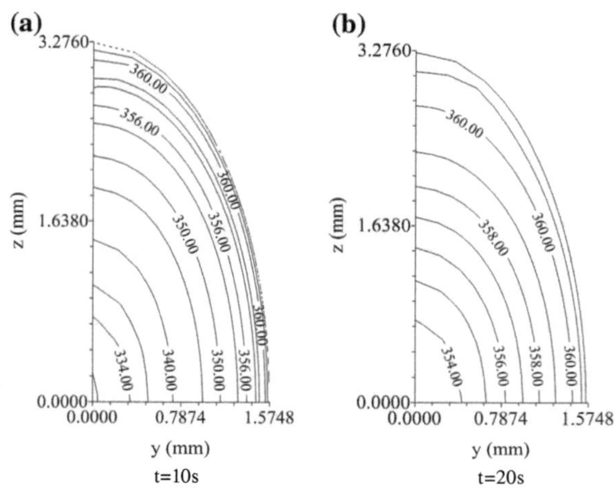

Fig. 12 Predicted temperature distribution (Kelvin) inside the wheat kernel during the drying at 87.8 °C (Model 1). **a** t = 10 s. **b** t = 20 s

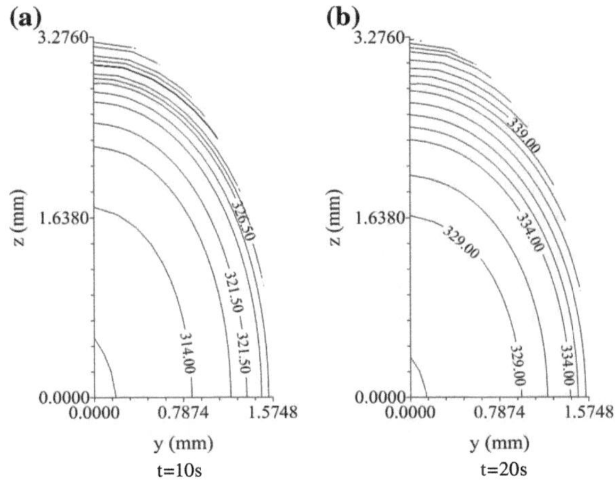

Fig. 13 Predicted temperature distribution (Kelvin) inside the wheat kernel during the drying at 87.8 °C (Model 2). **a** t = 10 s. **b** t = 20 s

Table 2 Liquid and vapor conductivities and fitting statistical data

Model	This work				Fortes et al. [28]		ERMQ ($\times 10^{-4}$)	$\bar{S}^2$ ($\times 10^{-3}$)
	Parameters							
	a_1	a_2	$h_m \times 10^{+6}$ (m/s)	h_c (W/m^{2}K)	a_1	a_2		
1	19.61	2.68	Infinity	Infinity	1	1	5.7484	1.5494
2	26.08	3.38	2.1	85.4467	1	1	1.40396	0.0054

4.3.3 Transport Coefficients

Table 2 presents the transport coefficients (kl and kv) as well as the residual error and variance obtained in the simulations. The small variance indicates that both the models present good agreement with the experimental data, specially model 2.

In general, the comparison between the transport coefficients reported in the literature is difficult due to the different models and calculation methods used, and also due to unlike composition and physical and chemical structure of the material. For an analysis, the values of the liquid and vapor conductivities of wheat-water system obtained in this work can be compared with those reported by Fortes et al. [28], in the drying conditions presented in Table 1. Fortes et al. [28] considers the wheat grain like sphere. As the Fortes and Okos's model was used in both works, the difference between the values may be attributed mainly to the factors: geometry assumptions; boundary conditions and additional convective terms in the formulation proposed here. Therefore of exposed, due to the good agreement obtained by comparison, we can say that the model is satisfactory to predict the liquid and vapor diffusion phenomenon in prolate spheroidal solid.

During the drying phenomenon, the grain shrinks by moisture removal and suffers dilation by heating. This modifies the shape of the body, affecting the area of heat and transfer and modifying the superficial roughness of the material. This last characteristic provides an increase of the turbulence level in the boundary layer, favoring this way the change of energy between the air and the material. These phenomena were not considered here. In addition, we can be said that although good fit have been obtained in this study, it is necessary to provide more attention to the quantitative study of superficial area and volume variations during the dehydration processes, especially in complexes situations, when multi directional deformations and temperature variations happen simultaneously.

5 Considerations About the Formulation

In this research, we develop generalized heat and mass transfer equations applicable for capillary-porous bodies with ellipsoidal configuration. Then, the model can be used in several situations such as:

(a) Process of wetting, drying, heating and cooling;
(b) Diffusion of liquid, vapor and heat in sphere, infinite cylinder and prolate spheroid with the same formulation. In this case it is necessary to change only dimension L_2, for fixed value of L_1;
(c) It is possible to treat diffusion problem with variable or constant thermo physical properties, equilibrium or convective boundary conditions. Changes in the model is unnecessary;
(d) Under numerical aspect it is possible to use uniform or non-uniform grid. Formulation reported in the cited references includes these procedures.
(e) No restriction about the nature of the solid (fruits, grains, vegetables, etc.) is required when it is considered as continuum media.

Limitation is related to chemical transformation and structure inside the solid and, moderate and high air-drying temperature (where effect of pressure must be considered). The structure of the porous bodies matrix may vary widely in shape (cells, fibers or graves, for example), and no uniform distribution of void is really verified. Besides many substances (mainly fruits and vegetables) undergo thermal decomposition when they are treated thermally [50]. Diffusion is the only moisture migration mechanism (capillarity, filtration and adsorption effects, for example, are not considered separately). Besides others important physical and chemical aspects such as shrinkage, dilation, browning, hardening, loss of volatile components are not considered.

6 Concluding Remarks

In this chapter, a mathematical modeling to predict heat transfer and moisture (liquid and vapor) migration in porous bodies with prolate spheroidal shape was developed. The finite-volume method was employed to solve the governing equations. Formulation has been applied to predicted wheat grain drying process. From the analysis of the results, we can conclude that:

(a) The model and the technical used has both great potential, being accurate and efficient to simulate many practical problems of diffusion such as heating, cooling, wetting and drying in solids.
(b) The equilibrium moisture content gradient can be considered the main thermodynamic force responsible for moisture transport inside the solid.
(c) It is very important to consider the real shape of the body to study diffusion process.
(d) Highest moisture and temperature gradients inside the ellipsoid of revolution are found near the surface and around the focal point, being these areas more susceptible to deformations, cracks and fissures that contribute to quality reductions of the wet porous solid.
(e) At the surface of the solid and for lower moisture content of wheat, vapor flux is the dominant migration mechanism. For high moisture content, both liquid and vapor fluxes are important.
(f) The satisfactory validation of the coupled models describing the drying phenomena in porous bodies with ellipsoidal geometry and applied to wheat grain has allowed the estimative of the thermo-physical parameters (liquid and vapor conductivities) for wheat with good accuracy.
(g) This work shows the facilities of the methodology to be applied for different spheroidal bodies, by changing the aspect ratio (L_2/L_1) only, when compared with the solution of the diffusion problem in sphere and cylinder where in these last cases, it is necessary diffusion equation applied to each specified geometry.

Acknowledgments The authors would like to express their thanks to CNPq (Conselho Nacional de Desenvolvimento Científico e Tecnológico, Brazil), CAPES (Coordenação de Aperfeiçoamento de Pessoal de Nível Superior, Brazil), and FINEP (Financiadora de Estudos e Projetos, Brazil) for supporting this work; to the authors of the references in this chapter that helped in our understanding of this complex subject, and to the Editors by the opportunity given to present our research in this book.

References

1. Fortes, M.: A non-equilibrium thermodynamics approach to transport phenomena in capillary-porous media with special reference to drying of grains and foods. Doctorate thesis, Faculty of Purdue University, United States of America (1978)
2. Middleman, S.: An Introduction to Mass and Heat Transfer: Principles of Analysis and Design. John Wiley & Sons Inc., New York (1998)

3. Crank, J.: The Mathematics of Diffusion. Oxford Science Publications, New York (1992)
4. Gebhart, B.: Heat Conduction and Mass Diffusion. McGraw-Hill, New York (1993)
5. Skelland, A.H.P.: Diffusional Mass Transfer. Wiley, New York (1974)
6. Lima, A.G.B., Nebra, S.A.: Theoretical analysis of the diffusion process inside prolate spheroidal solids. Drying Technol. **18**(1–2), 21–48 (2000)
7. Carmo, J.E.F.: Non steady-state diffusion phenomenon in oblate spheroidal solids. Case studies: drying of lentil. Doctorate thesis, Process Engineering, Federal University of Campina Grande, Campina Grande, Brazil (2004). (In Portuguese)
8. Lima, A.G.B., Queiroz, M.R., Nebra, S.A.: Simultaneous moisture transport and shrinkage during drying of solids with ellipsoidal configuration. Chem. Eng. J. **86**, 85–93 (2002)
9. Lima, A.G.B., Queiroz, M.R., Nebra, S.A.: Heat and mass transfer model including shrinkage applied to ellipsoidal products: case study for drying of bananas. Develop. Chem. Eng. Miner. Process. **10**, 281–304 (2002)
10. Oliveira, V.A.B., Lima, A.G.B.: Mass diffusion inside prolate spherical solids: An analytical solution. Braz. J. Agro-Ind. Prod. **4**(1), 41–50 (2002)
11. Lima, D.R., Farias, S.N., Lima, A.B.G.: Mass transport in spheroids using the Galerkin method. Braz. J. Chem. Eng. **21**(4), 667–680 (2004)
12. Carmo, J.E.F., Lima, A.G.B.: Drying of lentil including shrinkage: a numerical study. Drying Technol. **23**, 1977–1992 (2005)
13. Cihan, A., Kahveci, K., Hacihafizoğlu, O., Lima, A.G.B.: A diffusion based model for intermittent drying of rough rice. Heat Mass Transf. **44**, 905–911 (2008)
14. Hacihafizoğlu, O., Cihan, A., Kahveci, K., Lima, A.G.B.: A liquid diffusion model for thin-layer drying of rough rice. Eur. Food Res. Tech. **226**, 787–793 (2008)
15. Oliveira, V.A.B., Lima, A.G.B.: Drying of wheat based on the non-equilibrium thermodynamics: a numerical study. Drying Technol. **27**, 306–313 (2009)
16. Oliveira, V.A.B., Lima, W.C.P.B., Farias Neto, S.R., Lima, A.G.B.: Heat and mass diffusion and shrinkage in prolate spheroidal bodies based on non-equilibrium thermodynamics: a numerical investigation. J. Porous Media **14**(7), 593–605 (2011)
17. Oliveira, V.A.B., de Lima, A.G.B., Silva, C. J.: Drying of wheat: a numerical study based on the non-equilibrium thermodynamics. Int. J. Food Eng. **8**(3), Article 19 (2012)
18. Luikov, A.V., Mikhailov, Y.A.: Theory of Energy and Mass Transfer. Pergamon Press Ltd, Oxford (1965)
19. Luikov, A.V.: Heat and Mass Transfer in Capillary Porous Bodies. Pergamon Press, New York (1965)
20. Mikhailov, M.D., Shishedjiev, B.K.: Temperature and moisture distributions during contact drying of a moist porous sheet. Int. J. Heat Mass Transf. **18**, 15–24 (1975)
21. Luikov, A.V.: Systems of differential equations of heat and mass transfer in capillary porous bodies: review. Int. J. Heat Mass Transf. **18**, 1–14 (1975)
22. Fortes, M., Okos, M.R.: Drying theories: their bases and limitations as applied to foods and grains. In: Mujumdar, A. (ed.) Advances in Drying, vol. 1, pp. 119–154. Hemisphere Publishing Corporation, New York (1980)
23. Whitaker, S.: Heat and mass transfer in granular porous media. In: Mujumdar, A. (ed.) Advances in Drying, vol. 1, pp. 23-61. Hemisphere Publishing Corporation, New York (1980)
24. Fortes, M., Okos, M.R.: Non-equilibrium thermodynamics approach to heat and mass transfer in corn kernels. Trans. ASAE **24**, 761–769 (1981)
25. Ribeiro, J.W., Cotta, R.M., Mikhailov, M.D.: Integral transform solution of Luikov's equations for heat and mass transfer in capillary porous media. Int. J. Heat Mass Trans. **36**(18), 4467–4475 (1993)
26. Irudayara, J., Wu, Y.: Analysis and application of Luikov's heat, mass and pressure transfer model to a capillary porous media. Drying Technol. **14**(3–4), 803–824 (1996)
27. Santos, G.T., Fortes, M., Martins, J.H., Monteiro, P.M.B.: Convective drying analysis of a single wheat kernel based on an irreversible thermodynamic model. Drying Technol. **31**(16), 1979–1993 (2013)

28. Fortes, M., Okos, M.R., Barret Jr, J.R.: Heat and mass transfer analysis of intra-kernel wheat drying and rewetting. J. Agric. Eng. Res. **26**, 109–125 (1981)
29. Lima, A.G.B., Farias Neto, S.R., Silva, W.P.: Heat and mass transfer in porous materials with complex geometry: fundamentals and applications. In: Delgado, J.M.P.Q. (ed.) Heat and Mass Transfer in Porous Media. Springer, Berlin (2012)
30. Magnus, W., Oberhettinger, F., Soni, R.P.: Formulas and Theorems for the Special Functions of Mathematical Physics. Springer, Berlin (1966)
31. Abramowitz, M., Stegun, I.A.: Handbook of Mathematical Functions. Dover Publications Inc., New York (1972)
32. Philip, J.R., De Vries, D.A.: Moisture movement in porous materials under temperature gradients. Trans. Am. Geophys. Union **38**(2), 222–232 (1957)
33. De Vries, D.A.: Simultaneous transfer of heat and moisture in porous media. Trans. Am. Geophys. Union **39**(5), 909–916 (1958)
34. Krischer, O.: Die Wissenschaftlichen Grundlagen der Trocknungstechnik, vol. Kap. IX. Springer, Berlin (1963)
35. Berger, D., Pei, D.C.T.: Drying of hygroscopic capillary porous solids: a theoretical approach. Int. J. Heat Mass Transf. **16**, 293–302 (1973)
36. De Groot, S.R.: Thermodynamics of Irreversible Processes. North Holland Publishing Company, Amsterdan (1951)
37. Prigogine, I.: Thermodynamic of Irreversible Processes, 2nd edn. Wiley, New York (1961)
38. Oliveira, V.B.: Heat and mass transfer inside prolate spheroidal solids via thermodynamics of irreversible processes. Doctorate thesis, Process Engineering, Federal University of Campina Grande, Campina Grande, PB, Brazil (2004). (In Portuguese)
39. Brebbia, C.A., Dominguez, J.: Boundary Elements: an Introductory Course. McGraw-Hill Company, New York (1989)
40. Patankar, S.V.: Numerical Heat Transfer and Fluid Flow. Hemisphere Publishing Corporation, New York (1980)
41. Maliska, C.R.: Computational Heat Transfer and Fluid Mechanics, 2nd edn. LTC, Rio de Janeiro (2004). (In Portuguese)
42. Kwon, Y.W., Bang, H.: The finite Element Method Using Matlab. CRC Press, Bora Raton (1997)
43. Mohapatra, D., Rao, P.S.: A Thin layer drying model of parboiled wheat. J. Food Eng. **66**, 513–518 (2005)
44. Singh, H., Singh, A.K., Kushwaha, H.L., Singh, A.: Energy consumption pattern of wheat production in India. Energy **32**, 1848–1854 (2007)
45. FAO Statistical Yearbook 2013. World food and agriculture. Food and Agriculture Organization of the United Nations, Rome (2013)
46. Brooker, D.B., Bakker-Arkema, F.W., Hall, C.W.: Drying and Storage of Grains and Oilseeds. AVI Book, New York (1992)
47. Kazarian, E.A., Hall, C.W.: Thermal properties of grains. Trans. ASAE. **8**, 33–37 (1965)
48. Figliola, R.S., Beasley, D.E.: Theory and Design for Mechanical Measurement. Wiley, New York (1995)
49. Hartley, J.G.: Coupled heat and moisture transfer in soils: a review. In: Mujumdar, A.S. (ed.) Advances in Drying, vol. 4, pp. 199–247. Hemisphere Publishing Corporation, Washington (1987)
50. Shukla, K.N.: Diffusion Process During Drying of Solids. World Scientific, Singapore (1990)

GBI Method: A Powerful Technique to Study Drying of Complex Shape Solids

A. G. Barbosa de Lima, J. M. P. Q. Delgado, I. B. Santos, J. P. Silva Santos, E. S. Barbosa and C. Joaquina e Silva

Abstract This chapter briefly focuses on the theory and applications of drying process (heat and mass transfer) with particular reference to arbitrarily-shaped wet capillary-porous bodies. Here, a modeling based on the liquid diffusion theory and the mathematical formalism to obtain the exact solution of the governing equation via Galerkin-based integral method are presented. The model considers constant thermo-physical properties and convective boundary conditions at the surface of the solid. Applications have been done to different solids of revolution and wheat grain. Predicted results of the average moisture content, average temperature, and moisture content and temperature distributions within the porous solids are presented and discussed, and for particular situations they are compared with experimental drying data.

Keywords Drying · Analytical · Wheat · Prolate spheroid

A. G. Barbosa de Lima (✉) · J. P. Silva Santos · E. S. Barbosa · C. Joaquina e Silva
Department of Mechanical Engineering, Federal University of Campina Grande,
Av. Aprígio Veloso, 882, Bodocongó, Campina Grande, PB 58429-900, Brazil
e-mail: gilson@dem.ufcg.edu.br

J. P. Silva Santos
e-mail: jplitys@bol.com.br

E. S. Barbosa
e-mail: enivaldo@dem.ufcg.edu.br

C. Joaquina e Silva
e-mail: carlota.jsilva@gmail.com

J. M. P. Q. Delgado
LFC—Building Physics Laboratory, Civil Engineering Department, Faculty of Engineering,
University of Porto, Porto, Portugal
e-mail: jdelgado@fe.up.pt

I. B. Santos
Department of Physics, State University of Paraiba, Campina Grande, PB, Brazil
e-mail: ivoneetebs@gmail.com

J. M. P. Q. Delgado and A. G. Barbosa de Lima (eds.), *Transport Phenomena and Drying of Solids and Particulate Materials*, Advanced Structured Materials 48,
DOI: 10.1007/978-3-319-04054-7_2,

1 Introduction

Drying is a processing operation that play important role in many industries such as food, pharmaceutical, ceramics, chemical and waste treatment. Drying is the removal of unbounded water molecules in the liquid and/or vapor phases caused by a partial pressure difference of water vapor between the surface of the product to be dried and the air surrounding. It is the process most often used for quality assurance and stability during storage of biological products such as fruits, vegetables and grains, considering that reduced amount of water in the material reduce both biological activity and deterioration caused by microorganisms, and chemical and physical changes that occur during storage [1]. In the manufacturing process of ceramics products, clay minerals, when mixed with water, become highly plastic (hydroplasticity) and it can be molded easily without cracking. These physical and chemical characteristics maintain the body shape during handling and drying. The water added to ceramic clay cover the surface of the clay particle, fills small capillaries and porous, and causes a separation of particles [2]. After plastic forming and casting, products must be dried, in order, to removal of the moisture prior for further processing and/or firing. If this moisture content is not removed, the extreme temperature in the kiln will force out this water during firing, causing cracking and until explosion of the product.

Drying is a phenomenon that occurs with simultaneous heat and mass transport and volume variation (shrinkage by moisture removal and expansion by heating). During the drying process appear strong thermo-hydro-mechanical stresses which are caused by the emergence of moisture and temperature gradients within the body, mainly when the solid has a geometry that disfavor its mechanical strength as for example tips. Then, to study drying of solids it is important to control and optimize the process, avoiding damage to the material.

The heat and mass transport within the porous solids had been subject of the investigation for many years, because of their wide variety of industrial applications involving drying and wetting, and scientific interest. Because of the complexities of the mechanism involved in the process through the irregular void configuration within the porous bodies many theories has been proposed to predict moisture movements into the porous solids. For instance, liquid diffusion theory, capillary theory, evaporation-condensation theory, Luikov's theory, Philip and De Vries'theory, Krischer's theory, Berger and Pei's theory, and Fortes and Okos theory. In the hygroscopic porous bodies the pores are partially filled with liquid water and partially filled with non condensable gas and water vapor mixture. Thus, moisture can be transferred both in liquid and in gaseous phases, and the following possible mechanisms for moisture migration in a porous solid are cited in the literature: liquid diffusion, vapor diffusion, surface diffusion, effusion flow, thermodiffusion, transport by capillary forces, transport by gravity forces, transport by osmotic pressure, transport due to pressure gradients, and transport due to shrinkage [1, 3–5]. The liquid diffusion theory has gained the preference of the researchers. It considers that the water inside the solid migrates to surface only in the liquid phase. This assumption facilitates the analysis of the physical problem.

The theoretical treatment to predict moisture removal by drying formally is based upon the traditional method and start with a potential balance in a differential control volume with arbitrary shape, in macroscopic scale. The geometry of the body is one of the factors that are used to establish certain hypotheses in the description of a physical process. When liquid diffusion theory is used to predict drying process, the Fick's second law and the Fourier's law has been utilized to predict mass diffusion and heat conduction inside the solid, respectively.

On the basis of many researches in early works, the heat and mass diffusion within a capillary-porous body can be predicted solving the governing equations by mean of different techniques such as: numerical (finite-difference, finite-element, boundary-element and finite-volume), analytical (separation of variables, Laplace transform and Galerkin-based integral method) and semi-analytical (Generalized integral transform). For several cases, appropriate considerations and boundary conditions, and often constant thermo-physical properties for the medium are assumed.

In the literature are cited several works which use common geometry shaped bodies, such as plates, cylinders and spheres [6–9]. Thus, different numerical studies in oblate or prolate spheroidal solids are reported in the literature [10–24]. These formal solutions of the diffusion equation has been obtained for various boundary conditions with constant or variable diffusion coefficient, in homogeneous or heterogeneous and isotropic or anisotropic bodies, and in steady or unsteady cases. However, when we consider one- or two-dimensional geometries for describe drying of solids with arbitrary shape discrepancies in the drying and heating kinetics and distribution of moisture and temperature within the solid are found when compared with experimental results.

For obtaining the analytical solution of the governing transport equation is a difficult task especially when the solid has an irregular shape. Despite of this disadvantage, the exact solution of the diffusion equation applied to bodies with complex geometry and constant or convective boundary conditions are reported in the literature [25–40]. These mentioned works are directed to study uncoupled heat and/or mass transfer.

Since, rigorous solutions to the heat and mass diffusion in different geometrical shapes are essential, for the predictions of both the performance and control of the processes like heating, cooling, wetting and drying, for examples, so, we notice that there is a lack of studies that must take into account the heat and mass transfer in bodies with complex geometric shape, which is necessary in order to describe more adequately the drying process.

In this sense, this chapter aims to develop a mathematical model and its analytical solution to predict drying of solids of revolution via Galerkin-based integral method. Applications have been done to different solids of revolution and wheat grain (prolate spheroid).

2 Heat and Mass Transfer Modeling Applied to Irregularly-Shaped Solids

The complexity of real processes taking place in a capillary-porous materials not only leads to uncertainties and difficulties with their mathematical formulation, but also cause considerable problem with the numerical and analytical treatment of resulting equation. According to Farias et al. [41], in the macroscopic scale, diffusion modeling approach presents like advantages: less mathematics than the microscopic transport analysis, external effects can be incorporated in boundary conditions, easy to obtain analytical solution for some particular cases. However, like disadvantages we can cite: many questionable theoretical assumptions are required, extrapolation to areas outside region of experimental data are questionable [42].

A realistic approach of the physical-mathematical model for the drying process of a solid, it can be influenced by internal and external conditions and mechanism of moisture migration and heat flux inside the material.

From the transport general equation applied to a infinitesimal control volume, when source and convective terms are nulls, we obtain the diffusion equation in the short form expressed as:

$$\frac{\partial(\lambda\Phi)}{\partial t} = \nabla \cdot \left(\Gamma^{\Phi}\nabla\Phi\right) \tag{1}$$

The solution of the diffusion problems for various physical situations of interest often requires the need to establish certain assumptions in describing the physical process. One of them is related to the geometry of the body in which occurs the transport of matter or energy. Here, in order to enable the solution of the physical issue, the following considerations were taken:

(a) The solid is homogeneous and isotropic;
(b) The distribution of the potential Φ inside the solid is uniform at the beginning of the process;
(c) Thermophysical properties are constant throughout the process;
(d) The solid consists of dry substance and water in the liquid phase;
(e) The phenomenon of drying occurs by diffusion of liquid water inside the solid and by evaporation of water at the surface.

Considering the transport coefficients Γ^{Φ} and λ constant, the formal solution of Eq. (1) can be written as follows [27]:

$$\Phi = \sum_{n=1}^{N} C_n \psi_n e^{-\gamma_n t} + \Phi_e \tag{2}$$

whereas the C_n, γ_n and Φ_e are constants, and Ψ_n is a function independent of time. Putting Eq. (2) in (1) we obtain:

$$\frac{\partial}{\partial t}\left[\lambda \sum_{n=1}^{N} C_n \psi_n e^{-\gamma_n t} + \Phi_e\right] = \nabla \cdot \left\{\nabla\left[\Gamma^{\Phi}\left(\sum_{n=1}^{N} C_n \psi_n e^{-\gamma_n t} + \Phi_e\right)\right]\right\} \tag{3}$$

Developing the derivative of the Eq. (3) we obtain:

$$\lambda \sum_{n=1}^{N} C_n \psi_n (-\gamma_n) e^{-\gamma_n t} = \sum_{n=1}^{N} C_n e^{-\gamma_n t} \nabla \cdot \nabla(\Gamma^{\Phi} \psi_n) \tag{4}$$

By organizing Eq. (4) it can be obtained:

$$\sum_{n=1}^{N} \left[\nabla \cdot \left[\Gamma^{\Phi} \nabla \psi_n\right] + \lambda \gamma_n \psi_n\right] = 0 \tag{5}$$

The specific function Ψ_n is obtained as a linear combination of a set of basis function, f_j, [27] as follows:

$$\psi_n = \sum_{j=1}^{N} d_{nj} f_j \tag{6}$$

where d_{nj} are constants which must be determined.

Substituting Eq. (6) into (5) and applying the Galerkin method which consists in multiplying both members of Eq. (5) by $f_i dV$ and integrate over the volume of the solid we obtain [43]:

$$\sum_{j=1}^{N} d_{nj} \left[\frac{1}{V} \int_V f_i \nabla \cdot (\Gamma^{\Phi} \nabla f_j) dV + \lambda \gamma_n \frac{1}{V} \int_V f_i f_j dV\right] = 0 \tag{7}$$

In matrix form Eq. (7) can be rewrite as follows:

$$(\overline{A} + \gamma_n \overline{B}) \overline{d}_n = 0 \tag{8}$$

where $\bar{A}$ and $\bar{B}$ are square matrices of N×N elements, whose elements are calculated by the following equations:

$$a_{ij} = \left[\frac{1}{V} \int_V f_i \nabla \cdot (\Gamma^{\Phi} \nabla f_j) dV\right] \tag{9}$$

$$b_{ij} = \frac{1}{V} \int_V \lambda f_i f_j dV \tag{10}$$

Coefficients $d_{n1}, d_{n2}, \ldots, d_{nN}$ in Eq. (6) are elements of vector $\bar{d}_n$ in Eq. (8). It can be observed also that matrix $\overline{B}$ is symmetrical, so $b_{ij} = b_{ji}$. Matrix $\overline{A}$ is symmetrical as well.

Since the linear equations originating in Eq. (8) are homogeneous, $\gamma_1, \gamma_2, \ldots, \gamma_N$ can be obtained to make the determinant of the matrix $\left(\overline{A} + \gamma\overline{B}\right)$ equal to zero.

Once the eigenvalues, γ_n, have been determined the values of coefficients d_{nj} corresponding to each γ_n can be obtained. Further, since the simultaneous equations resulting from Eq. (8) are homogeneous, one of the coefficients d_{nj} can be selected arbitrarily to be equal to 1 without any loss of generality. Therefore, for a specified d_{nj}, a system of (N−1) equations should be solved to obtain $d_{n2}, d_{n3}, \ldots, d_{nN}$.

The diffusion equation relates time and space variations of the potential Φ; it governs the transfer process within a body. For determine the distribution of Φ inside a porous body at any instant, i.e., to solve the diffusion equation, we necessitate to know the distribution of Φ at the initial instant (initial conditions), the geometry of the body, and the law of interaction between the surrounding medium and the body surface (boundary condition). Thus, we have a well-posed mathematical model.

In this research we used the identity, Eq. (11) to apply the boundary conditions,

$$\int_V f_i \nabla \cdot \left(\Gamma^{\Phi} \nabla f_j\right) dV = \int_V \nabla \cdot \left(\Gamma^{\Phi} f_i \nabla f_j\right) dV - \int_V \Gamma^{\Phi} \nabla f_i \cdot \nabla f_j dV \tag{11}$$

Since Γ^{Φ} is constant, Eq. (11) can be written as follows:

$$\int_V f_i \nabla \cdot \left(\Gamma^{\Phi} \nabla f_j\right) dV = \int_S \Gamma^{\Phi} f_i \nabla f_j \vec{n} \cdot d\vec{S} - \int_V \Gamma^{\Phi} \nabla f_i \cdot \nabla f_j dV \tag{12}$$

or yet,

$$\int_V f_i \nabla \cdot \left(\Gamma^{\Phi} \nabla f_j\right) dV = \int_S \Gamma^{\Phi} f_i \left(\frac{\partial f_j}{\partial n}\right) dS - \int_V \Gamma^{\Phi} \nabla f_i \cdot \nabla f_j dV \tag{13}$$

For analysis consider the following boundary condition, which is a specification of a linear combination of the values of a function Φ and the values of its derivative on the boundary of the domain:

$$-\Gamma^{\Phi} \frac{\partial \Phi}{\partial \vec{n}} = h(\Phi - \Phi_e) \tag{14}$$

where h represent the convective transfer coefficient and $\vec{n}$ is the normal vector to surface. If h tends to an infinity value we have a so called Dirichlet condition

(Φ prescribed). When h tend to zero we have a so called Newmann condition (flux prescribed), and for another situation we have a so called Robin condition (convective boundary condition).

Appling Eq. (2) in the (14) it is possible to prove that:

$$-\Gamma^{\Phi}\frac{\partial f_j}{\partial \vec{n}} = hf_j \tag{15}$$

For boundary conditions of the 1st kind (Φ prescribed) $f_j = 0$. For boundary conditions of the 2nd kind (flux prescribed) we have that $\partial f_j/\partial n = 0$ already for the boundary conditions of the 3rd kind (convective boundary condition), we have $-\Gamma^{\Phi}\partial f_j/\partial n = hf_j$.

For obtain the constants C_n of Eq. (2) we use the condition at $t = 0$, i.e., $\Phi = \Phi_0$. Then, applying this value in Eq. (2) we obtain the following equation:

$$\Phi_0 = \sum_{n=1}^{N} C_n\Psi_n + \Phi_e \tag{16}$$

Multiplying Eq. (16) by f_i dV and integrating over the volume we obtain [43]:

$$\int_V f_i(\Phi_0 - \Phi_e)dV = \int_V f_i \sum_{n=1}^{N} C_n\Psi_n dV \tag{17}$$

The solution of Eq. (17) will be a set of N linear algebraic equations, which allow determine the parameters C_n, completing the solution of the physical-mathematical problem.

Function f_j is called the Galerkin function and it is obtained by the multiplication of function of position φ by an element of a complete set of functions. Function φ is selected satisfying the homogeneous boundary condition. Function f_j with j varying from 1 to N constitutes a set of base functions.

The method to select basis functions for boundary conditions of the first kind, $f_j^{(1)}$, which correspond to the equilibrium boundary condition at the surface of the body, is given in the literature [36, 42].

The basis functions of first kind are determined from the first basis function, which is given by:

$$f_1 = \varphi_1\varphi_2\varphi_3\ldots\varphi_m \tag{18}$$

where $\varphi_m = 0$ is an equation that describes one of the surfaces of the body, and m is the number of surfaces that define the body to be studied.

Each base function should tend towards zero at the boundary of the solid. Some, but not all, of the basis functions can be zero at some point in the solid.

The basis functions of the second and third kind are defined by [36]:

$$f_j^{(2)} = f_j^{(1)}(\varphi_m H - 1) \tag{19}$$

where H is given by

$$H = \left. \frac{\nabla f_j^{(1)} \cdot \nabla \varphi_m}{f_j^{(1)} \cdot \nabla \varphi_m \cdot \nabla \varphi_m} \right|_{\varphi_m = 0} \tag{20}$$

and

$$f_j^{(3)} = f_j^{(2)} \left(\varphi_m H' - \frac{\Gamma^\Phi}{h} \right) \tag{21}$$

where H' can be determined by:

$$H' = \left. \frac{1}{\frac{\partial \varphi_m}{\partial \vec{n}}} \right|_{\varphi_m = 0} = \left. \frac{1}{\nabla \varphi_m \cdot \nabla \varphi_m} \right|_{\varphi_m = 0} \tag{22}$$

The average value of the potential Φ within the solid can be calculated as follows:

$$\overline{\Phi} = \frac{1}{V} \int_V \Phi dV \tag{23}$$

where V is the volume of the solid.

To describe mass transport within the solid with arbitrary geometry, it is considered in the Eq. (1), $\lambda = 1$, $\Gamma^\Phi = D$ (mass diffusion coefficient), and $\Phi = M$ (moisture content in dry basis), and $h = h_m$ (convective mass transfer coefficient), thus the mass diffusion equation in transient state without mass generation (Fick's law) will be written as follows:

$$\frac{\partial M}{\partial t} = \nabla \cdot (D \nabla M) \tag{24}$$

For heat transfer we consider in Eq. (1) $\Gamma^\Phi = k$ (thermal conductivity), $\lambda = \rho c_p$ (density and specific heat) and $\Phi = \theta$ (temperature), and $h = h_c$ (convective heat transfer coefficient), thus the heat conduction equation in transient state without energy generation (Fourier's law) will be written as follows:

$$\frac{\partial \theta}{\partial t} = \nabla \cdot (\alpha \nabla M) \tag{25}$$

where $\alpha = k/(\rho c_p)$ represents the thermal diffusivity.

3 Applications

3.1 Drying of Solids of Revolution

A solid of revolution is a solid figure obtained by rotating a plane curve in space around some straight line (the axis) that lies on the same plane (axis coplanar). It can be defined yet like a three-dimensional figure formed by revolving a plane area about a given axis. However the curve does not intercept the axis. Thus, considers a solid of revolution pictured in Fig. 1.

The contour of the solid of revolution represented in Fig. 1 is defined by:

$$\frac{r^m}{a^m} + \frac{z^2}{b^2} = 1 \tag{26}$$

where m is a number that defines the geometric shape of the solid being studied.

Thus, a set of basis functions of the 1st kind is defined for instance as follows:

$$f_j^{(1)} = \varphi_1 r^{m_j} z^{n_j} \tag{27}$$

where

$$\varphi_1 = 1 - \frac{r^m}{a^m} - \frac{z^2}{b^2} \tag{28}$$

being, $j = 0, 1, 2, 3, \ldots$, $m_j = 0, 2, 4, 6\ldots$ and $n_j = 0, 2, 4, 6\ldots$

The volume of the solid shown in Fig. 1 can be calculated by:

$$V = \int_0^{2\pi}\int_0^{a}\int_0^{b\sqrt{1-r^m/a^m}} r\,dz\,dr\,d\theta \tag{29}$$

Then, the coefficients presented in Eqs. (9), (10) and (17) can be given as follows:

$$a_{ij} = \int_0^{2\pi}\int_0^{a}\int_0^{b\sqrt{1-r^m/a^m}} f_i \nabla \cdot (\Gamma^{\Phi} \nabla f_j) r\,dz\,dr\,d\theta \tag{30}$$

$$b_{ij} = \int_0^{2\pi}\int_0^{a}\int_0^{b\sqrt{1-r^m/a^m}} f_i f_j r\,dz\,dr\,d\theta \tag{31}$$

$$\int_0^{2\pi}\int_0^{a}\int_0^{b\sqrt{1-r^m/a^m}} f_i(\Phi_0 - \Phi_e) r\,dz\,dr\,d\theta = \int_0^{2\pi}\int_0^{a}\int_0^{b\sqrt{1-r^m/a^m}} f_i \sum_{n=1}^{N} C_n \Psi_n r\,dz\,dr\,d\theta \tag{32}$$

Fig. 1 Geometrical representation of a solid of revolution

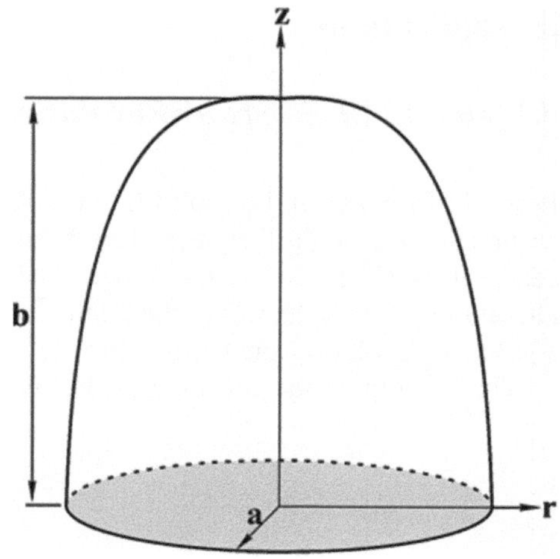

By varying the parameter m, in Eq. (26), and fixed the aspect ratio (b/a = 2.0), we generated the following solids of revolution presented in Fig. 2. Further, for fixed value of the parameters "m" and "a", for example, a = 1.0 and m = 2.0 and varying the value of the parameter "b" we obtain different ellipsoids of revolution. When the solid in which the axis of revolution is greater than the other axis (b > a), the solid is said to be prolate spheroid (Fig. 2c). For the other hand, if the axis of revolution is smaller than the other (b < a), it is called oblate spheroid. When the axis of revolution is equal to the other axis (b = a), so, the ellipsoid is known like sphere.

Figure 3 shows the dimensionless average moisture content as a function of mass transfer Fourier number (Dt/a^2) for different values of the parameter m and fixed aspect ratio b/a considering equilibrium boundary condition at the surface of the solid. It is clear from Fig. 3 that for the same Fourier number there are different moisture levels depending of the "m" parameter value, so factor m in Eq. (22) defines how the body shape influenced the drying kinetics. By examining this figure, one can observe that for m = 0.5, the dimensionless average moisture content within the solid decreases faster than for m = 4.0. In other words, for the same aspect ratio, the higher the value of m in Eq. (24) the lower the area/volume ratio will be formed. Since the drying velocity (drying rate) is directly proportional to the area/volume relationships, it is noticed that for m = 4.0, the drying process of this solid will be slower when compared to other solids of revolution studied. This observation is consistent with existing works in literature [14, 15, 35, 38, 44]. Higher rates of drying (high concentration gradients) and heating (high temperature gradients) induce elevated thermal, hydric and mechanical stresses within the solid, which could result in the manifestation of cracks or distortions in the product, which may come to compromise its final quality post-drying.

Figures 4 and 5 show the distribution of the dimensionless moisture content inside the solids of revolution with aspect ratio b/a = 2.0, in two dimensionless

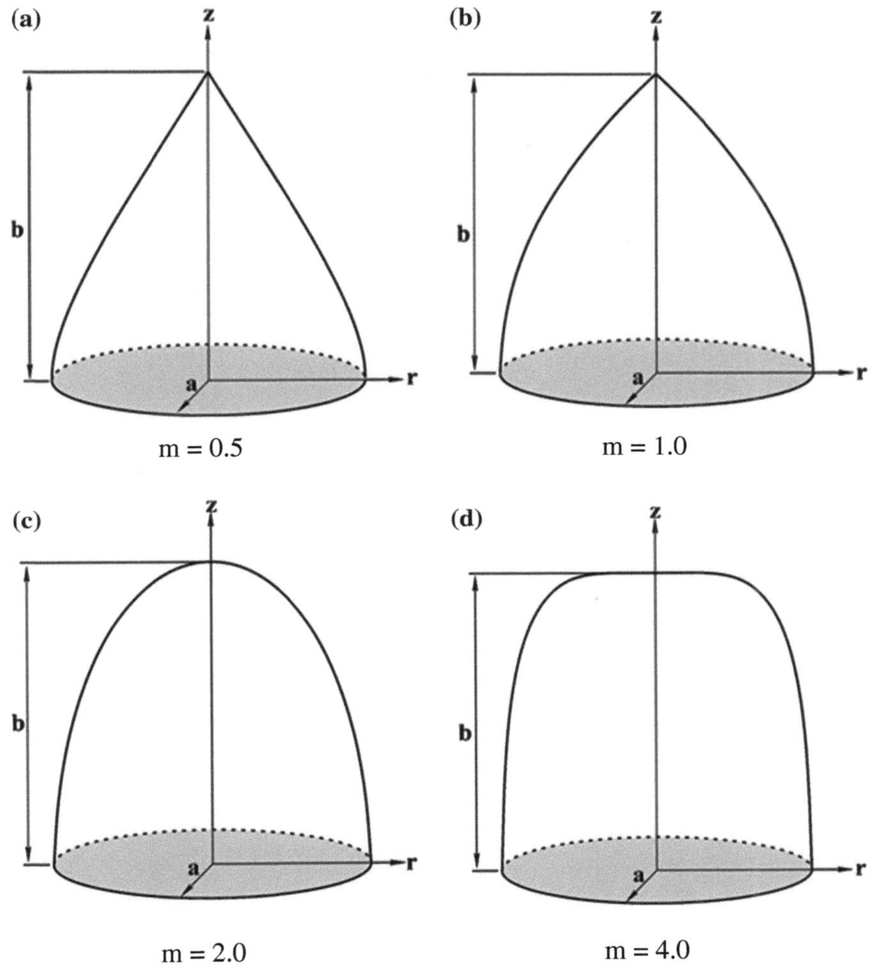

Fig. 2 Solids of revolution with different shapes, **a** m = 0.5, **b** m = 1.0, **c** m = 2.0, **d** m = 4.0

time (Fourier number, Fo = Dt/a^2): 0.0708617 and 0.141723. From these figures, one can verify the existence of concentration gradients during the drying process and that they are higher in early times for each of the solids. However, it is observed that these moisture gradients are higher in solids that have a lower "m" value, which represent the shape factor of the body, being consistent with Fig. 3. Since higher moisture gradients will result in large movement of fluid within the solid, it dries more quickly.

It is also worth pointing out that the region near the point z = 2.0 upholds the lowest moisture concentrations for all times in anyone of the solids. This phenomenon is related with the highest concentration gradients that has occurred at

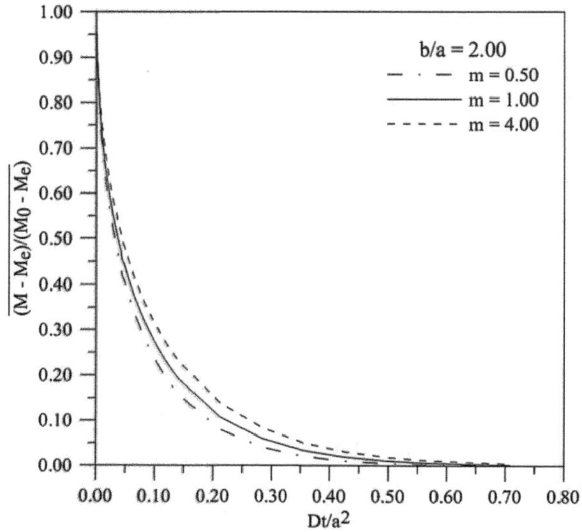

Fig. 3 Dimensionless average moisture content within three solids of revolution with the same aspect ratio as a function of Fourier number

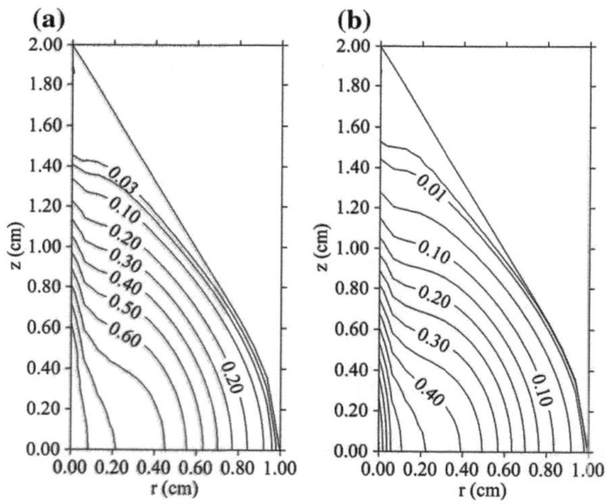

Fig. 4 Dimensionless moisture content distribution $[(M - M_e)/(M_0 - M_e)]$ within a solid of revolution formed with m = 0.5 and b/a = 2.0 for dimensionless times: **a** Fo = 0.0708617 and **b** Fo = 0.141723

the edge of the solid close to z = 2.0, being lower when the solid tend to a sphere (z = 1.0). Therefore, if the drying rate is such that it will cause a fracture or a deformation in the solid, it is very likely that such effects occur in the edge region. For this reason, several authors mention this edge effect [14, 15, 23, 24, 35, 44]. Further, one can observe the presence of iso-concentration lines, which have the same form of the solid. The moisture concentrations are higher in the center of the solid thus showing that the movement of liquid occurs from the interior to the surface of the solid of revolution, as expected.

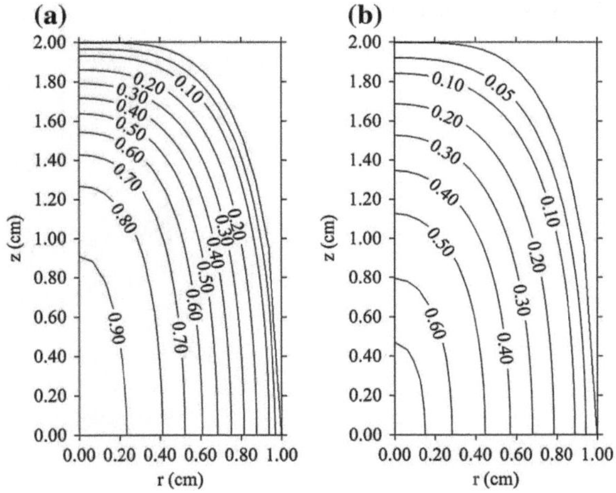

Fig. 5 Dimensionless moisture content distribution $[(M - M_e)/(M_0 - M_e)]$ within a solid of revolution formed with m = 4.0 and b/a = 2.0 for dimensionless times: **a** Fo = 0.0708617 and **b** Fo = 0.141723

3.2 Drying of Wheat Grain

For the theoretical study of drying considered in this topic we use a prolate spheroid (ellipsoid of revolution formed when m = 2.0) whose geometry is resembled to wheat grain, as illustrated in Fig. 6.

To determine the mass diffusion coefficient it was used the equation proposed by [45], as follows:

$$D = 0.543\bar{M}^{\left(-2.8554\times10^{-5}T+1.6432\right)}\mathrm{Exp}\left[(0.4113T - 30.2634)\bar{M} + (0.022776T - 9.7271)\right] \tag{33}$$

From the Eq. (33), the mass diffusion coefficient was obtained using the average value of moisture content as follows:

$$\bar{D} = \frac{1}{(M_0 - M_e)}\int_{M_e}^{M_0} D(\bar{M})d\bar{M} \tag{34}$$

For determination of the convective mass transfer coefficient was used the following relations reported by [45]:

$$h_m = \frac{D_{atm}}{RT_{abs}d_p}\left[2.0 + 0.6Re^{1/2}Sc^{1/3}\right] \tag{35}$$

where D_{atm} represents the molecular diffusion coefficient of water in air, Sc and Re represent the Reynolds and Schmidt numbers, R is the Universal gas constant (as applied to water vapour) and dp is the equivalent diameter.

The convective heat transfer coefficient was calculated as follows [45]:

Fig. 6 Wheat grain kernel

$$h_c = \frac{k_a}{d_p}\left[2.0 + 0.6Re^{1/2}Pr^{1/3}\right] \tag{36}$$

where Pr represents the Prandtl number.

The Reynolds, Schmidt and Prandtl numbers are given by:

$$Re = \frac{\rho v d_p}{\mu} \tag{37}$$

$$Sc = \frac{\mu}{\rho D_{atm}} \tag{38}$$

$$Pr = \frac{C_p \mu}{k} \tag{39}$$

In the Eqs. (37)–(39), ρ, v, μ, c_p and k represent density, velocity, viscosity, specific heat and thermal conductivity of the air, respectively. Equivalent diameter (d_p) was obtained considering the wheat grain as a prolate spheroid. Initially, the volume of the grain was calculated using Eq. (29). Following, this volume was assumed to be equal to the volume of a sphere of diameter d_p. Simulations were performed using Eqs. (24) and (25). Table 1 presents the data used in the simulations.

Figures 7 and 8 illustrate, respectively, the results of the dimensionless average moisture content and dimensionless center temperature obtained in this study compared with experimental drying data to wheat [45] considering convective boundary condition at the surface of the solid. Based on the approach taken for this research and observing the drying and heating kinetics, it becomes apparent that a good agreement was obtained. We notice that, the higher the mass diffusion coefficient, the lower the drying time, i.e., the drying process becomes faster, fixed the convective mass transfer coefficient. For a fixed mass diffusion coefficient

Table 1 Physical and geometrical parameters used in the simulations [14, 45, 46]

Drying air			Wheat grain					
T (°C)	RH (%)	v (m/s)	a (cm)	b (cm)	M_0 (d.b.)	M_e (d.b.)	θ_e (°C)	θ_0 (°C)
67.5	13.3	1.61	0.1575	0.3276	0.2560	0.0362	67.5	26.0
			k (W/mK)	ρ (kg/m^3)	c_p (J/kgK)	h_c (W/m^{2}K)	$h_m \times 10^{+6}$ (m/s)	$D \times 10^{+11}$ (m^2/s)
			0.1400	790.00	2223.76	82.942	1.03823	5.73003

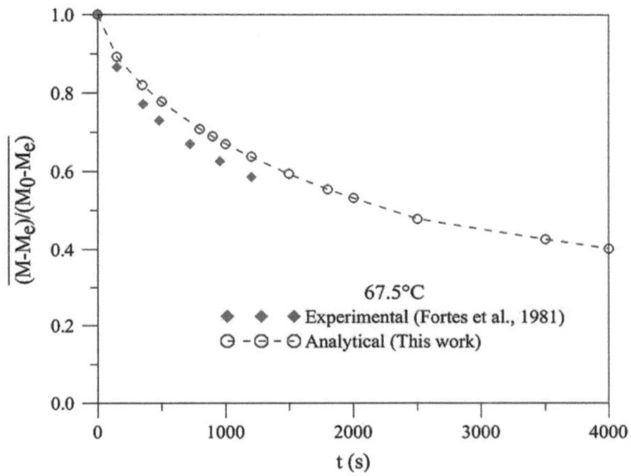

Fig. 7 Dimensionless average moisture content of the wheat grain as a function of time

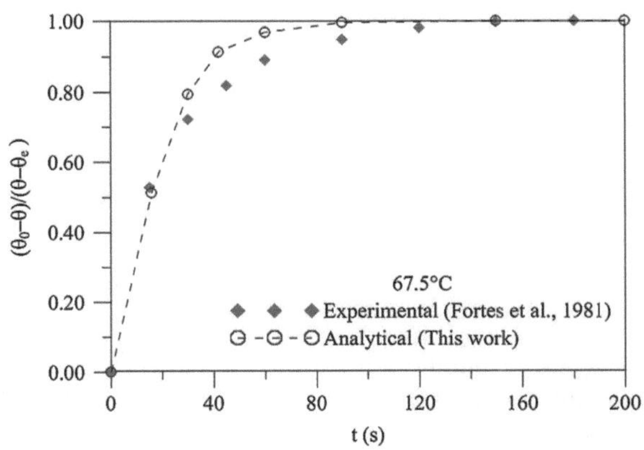

Fig. 8 Dimensionless temperature at the center of the wheat grain as a function of time

increasing the convective mass transfer coefficient we have increased drying rate, thus the solid dry faster.

Figures 9 and 10 illustrate the moisture content and temperature distributions inside the grain at two different elapsed times (1800 and 4000 s). Based on the analysis of this figure, we note the presence of concentration iso-lines, which has the same shape as prolate spheroid. It is also evident that the movement of liquid

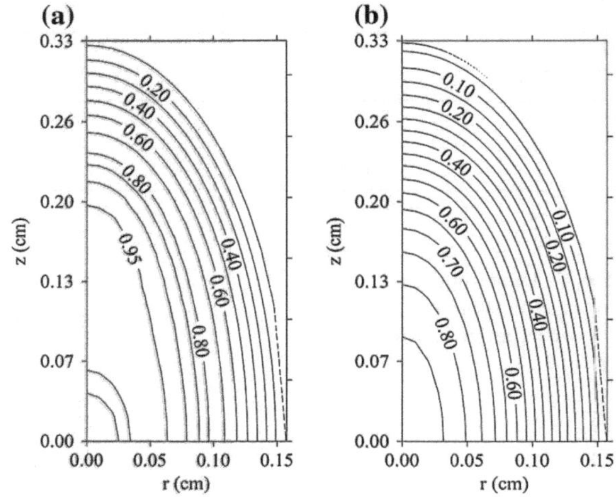

Fig. 9 Distribution of dimensionless moisture content during the wheat grain drying: **a** t = 1800 s and **b** t = 4000 s (T = 67.5 °C and RH = 13.3 %)

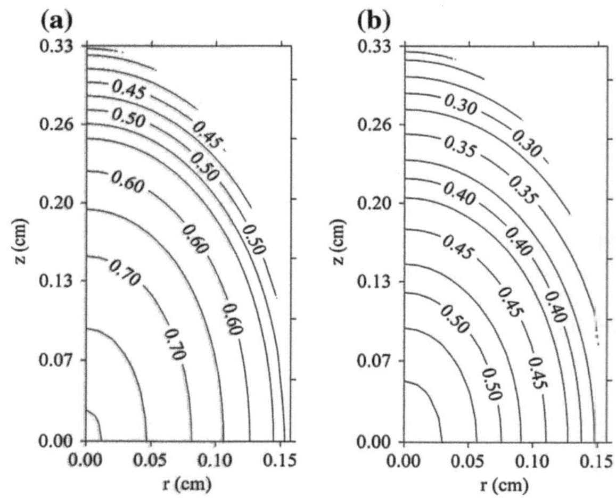

Fig. 10 Distribution of dimensionless temperature during the wheat grain drying: **a** t = 8 s and **b** t = 14 s (T = 67.5 °C and RH = 13.3 %)

takes place from the center of the wheat grain to the surface thereof meanwhile the heat flux occurred from the surface to center of the solid. The moisture content and temperature inside the wheat grain checked at all times occurred in a non-homogeneous form. Specifically for the instant 4000 s, it can be noted that the moisture content of the grain is close to reach the commercial moisture content, so, at this instant the drying is almost complete. The longer the drying time, the smaller the presence of moisture gradient inside the solid and thus it has a slower drying rate, tending to zero at the final of the drying, when the grain reach the equilibrium moisture content. Further, the time to grain reach the air temperature is very short, less than the 200 s, so, moisture migration has occurred at isothermal conditions after this time.

4 Concluding Remarks

Upon completion of this chapter and based on the drying data analysis, the following conclusions can be cited:

(a) The mathematical model and GBI technique used for obtaining the analytical solution was appropriate to describe drying problem. Thus it is also suitable for predicting transient problems such as cooling, heating or wetting. From the model solution, is possible to obtain the moisture content and temperature distributions within the solid as well as analyze the drying and heating kinetics.
(b) The body shape has a direct influence on the drying and heating kinetics, since this parameter is directly related to the area/volume of the body. For higher area/volume relationships the solid dry faster.
(c) During drying there was a difference in moisture content and temperature within the porous solid, so that the flux of moisture occurs from the center of the solid to the surface thereof, meanwhile the heat flux occurs from surface to center of the solid.
(d) It was found that the higher moisture and temperature gradients occur on the edge of the ellipsoid of revolution. This fact shows that this region is more affected by thermo-hydro-mechanical stresses, thus, being more susceptible to the appearance of defects such as cracks, fracture, warping and deformations.
(e) The diffusion coefficient affects the mass transfer process, reporting the velocity at which this process occurs. The greater the mass diffusivity the faster drying is found.

Acknowledgments The authors would like to express their thanks to CNPq (Conselho Nacional de Desenvolvimento Científico e Tecnológico, Brazil), CAPES (Coordenação de Aperfeiçoamento de Pessoal de Nível Superior, Brazil), and FINEP (Financiadora de Estudos e Projetos, Brazil) for supporting this work; to the authors of the references in this paper that helped in our understanding of this complex subject, and to the Editors by the opportunity given to present our research in this book.

References

1. Brooker, D.B., Bakker-Arkema, F.W., Hall, C.W.: Drying and storage of grains and oilseeds. AVI Book, New York (1992)
2. Brosnan, D.A., Robinson, G.C.: Introduction of drying of ceramics: with laboratory exercises The American Ceramic Society, Westerville (2003)
3. Strumillo, C., Kudra, T.: Drying: principles, science and design. Gordon and Breach Science Publishers, New York (1986)

4. Fortes, M.: A Non-equilibrium thermodynamics approach to transport phenomena in capillary-porous media with special reference to drying of grains and foods. Ph.D. thesis, Pardue University (1978)
5. Fortes, M., Okos, M.R.: Drying theories: their bases and limitations as applied to foods and grains. In: Mujumdar, A.S. (ed.) Advances in drying. Hemisphere Publishing Corporation, Washington (1980)
6. Crank, J.: The mathematics of diffusion. Oxford Science Publications, New York (1992)
7. Gebhart, B.: Heat conduction and mass diffusion. McGraw-Hill Inc, New York (1993)
8. Luikov, A.V.: Analytical heat diffusion theory. Academic Press, Inc., Ltd, London (1968)
9. Carslaw, H.S., Jaeger, J.C.: Conduction of heat in solids. Oxford University Press, New York (1959)
10. Elvira, C.: The diffusion process modelling in elliptic shaped bodies. In: Proceedings of the. 6th International Congress Engineering and Food, vol. 1, London (1990)
11. Haghighi, K., Irudayaraj, J., Stroshine, R.L., Sokhansanj, S.: Grain kernel drying simulation using the finite element method. Trans. ASAE **33**(6), 1957–1965 (1990)
12. Lu, R., Siebenmorgen, T.J.: Moisture diffusivity of long-grain in rice components. Trans. ASAE **35**(6), 1955–1961 (1992)
13. Sarker, N.N., Kunze, O.R., Stroubolis, T.: Finite element simulation of rough rice drying. Drying Technol. **12**(4), 761–775 (1994)
14. Lima, A.G.B.: Diffusion phenomena in solid prolate spheroid. Case study: Banana drying. Doctorate thesis. State University of Campinas, SP (1999)
15. Lima, A.G.B., Nebra, S.A.: Theoretical analysis of the diffusion process inside prolate spheroidal solids. Drying Technol. **18**(1–2), 21–48 (2000)
16. Lima, A.G.B., Queiroz, M.R., Nebra, S.A.: Heat and mass transfer model including shrinkage applied to ellipsoids products: case study: drying of bananas. Develop. Chem. Eng. Mineral Process. **10**(3–4), 281–304 (2002)
17. Lima, A.G.B., Nebra, S.A., Queiroz, M.R.: Simultaneous moisture transport and shrinkage during drying of solids with ellipsoidal configuration. Chem. Eng. J. **86**(1–2), 85–93 (2002)
18. Gastón, A.L., Abalone, R.M., Giner, S.A.: Wheat drying kinetics. Diffusivities for sphere and ellipsoid by finite elements. J. Food Eng. **52**, 313–322 (2002)
19. Carmo, J.E.F.: Diffusion Phenomena in oblate spheroidal solids: modelling and simulation. Master's thesis, Federal University of Paraiba, Campina Grande (2000) (In Portuguese)
20. Carmo, J.E.F., Lima, A.G.B.: Drying of lentil including shrinkage: a numerical simulation. Drying Technol. **23**, 1977–1992 (2005)
21. Igathinathane, C., Chattopadhyay, P.K.: Surface area of general ellipsoid shaped food materials by simplified regression equation method. J. Food Eng. **46**, 257–266 (2000)
22. Carmo, J.E.F., Lima, A.G.B.: Mass transfer inside oblate spheroidal solids: modelling and simulation. Braz. J. Chem. Eng. **25**(1), 19–26 (2008)
23. Oliveira, V.A.B., Lima, W.C.P.B., Farias Neto, S.R., Lima, A.G.B.: Heat and mass diffusion and shrinkage in prolate spheroidal bodies based on non-equilibrium thermodynamics: a numerical investigation. J. Porous Media **14**(7), 593–605 (2011)
24. Carmo, J.E.F., Lima, A.G.B., Joaquina e Silva, C.: Continuous and intermittent drying (tempering) of oblate spheroidal bodies: modeling and simulation. Int. J. Food Eng. **8**(3), 1–17 (2012) (Article 20)
25. Norminton, E.J., Blackwell, J.H.: Transient heat flow from constant temperature spheroids and the thin circular disk. Quart. J. Mech. Appl. Math **XVII**(1), 65–72 (1964)
26. Haji-Sheikh, A., Sparrow, E.M.: Transient heat conduction in a prolate spheroidal solid Trans. ASME. J. Heat Transf. **88**(3), 331–333 (1966)
27. Haji-Sheikh, A.: On solution of parabolic partial differential equations using Galerkin functions. In: Payne, F.R., Corduneanu, C.C., Haji-Sheikh, A., Huang, T. (eds.) Integral methods in science and engineering. Hemisphere Publishing Corporation, New York, USA (1986)
28. Haji-Sheikh, A., Lakshminarayanan, R.: Integral solution of diffusion equation: Part 2—boundary conditions of second and third kinds. J. Heat Transf. **109**(3), 557–562 (1987)

29. Oliveira, V.A.B., Lima, A.G.B., Joaquina e Silva, C.: Drying of wheat: a numerical study based on the non-equilibrium thermodynamics. Int. J. Food Eng. **8**(3), 1–21 (2012) (Article 19)
30. Alassar, R.S.: Heat conduction from spheroids. J. Heat Transf. **121**(2), 497–499 (1999)
31. Oliveira, V.A.B.: Diffusion in prolate spheroidal solids: an analytical solution, Master's thesis, Federal University of Paraíba, Campina Grande (2001) (In Portuguese)
32. Oliveira, V.A.B., Lima, A.G.B.: Unsteady state mass diffusion in prolate spheroidal solids: an analytical solution. In: 2nd Inter-American Drying Conference. Boca Del Rio, Vera Cruz, Mexico (2001)
33. Oliveira, V.A.B., Lima, A.G.B.: Mass diffusion inside prolate spherical solids: an analytical solution. Braz. J. Agro-ind. Prod. **4**(1), 41–50 (2002)
34. Farias, S.N.: Drying of spheroidal solids using the Galerkin method. Master's thesis, Federal University of Paraíba, Campina Grande, Brazil (2002) (In Portuguese)
35. Lima, D.R., Farias, S.N., Lima, A.G.B.: Mass transport in spheroids using the Galerkin method. Braz. J. Chem. Eng. **21**(1), 667–680 (2004)
36. Beck, J.V., Cole, K.D., Haji-Sheikh, A., Litkouhi, B.: Heat conduction using green's functions. Hemispheric Publishing Corporation, New York (1992)
37. Hacihafizoglu, O., Cihan, A., Kahveci, K., Lima, A.G.B.: A liquid diffusion model for thin-layer drying of rough rice. European Food Res. Technol. **226**(4), 787–793 (2008)
38. Santos, I.B., Silva, L.P.L., and Lima, A.G.B.: Mass transport in solids with arbitrary shape via Galerkin-based integral method using convective boundary condition. In: 9th Argentinean Congress on Computational Mechanics and 31th Iberian-Latin-American Congress on Computational Methods in Engineering, vol. 1, pp. 2865–2881, Buenos Aires, Argentina (2010)
39. Silva, A.A., Santos, I.B., Lima, A.G.B.: Mass transport in solids with arbitrary shape via GBI method: an analytical study. In: 6th National Congress of Mechanical Engineering, vol. 1, Campina Grande, Brazil (2010)
40. Santos, I.B., Silva, L.P.L., Lima, A.G.B.: Mass transfer in irregularly-shaped solid: an exact solution using the galerkin-based integral method. Def. Diff. Forum **326–328**, 199–204 (2012)
41. Farias Neto, S.R., Farias, F.P.M., Delgado, J.M.P.Q., Lima, A.G.B., Cunha, A.L.: Cyclone: Their characteristics and drying technological applications. In: Delgado, J.M.P.Q. (ed.) Industrial and Technological Applications of Transport in Porous Materials, vol. 36, pp. 1–36 Springer, Heidelberg (2013)
42. Rosen, H.N.: Recent advances in the drying of solid wood. In: Mujumdar, A.S. (ed.) Advances in Drying, vol. 4, Hemisphere Publishing Corporation, Berlin (1987)
43. Kantorovich, L.V., Krylov, V.I.: Approximate methods of higher analysis. Advanced Calculus. Wiley, New York (1960)
44. Nascimento, J.J.S., Transient diffusion phenomenon in parallelepipeds solids. Case studied: Drying of ceramic materials. Ph.D. thesis, Federal University of Paraíba, João Pessoa (2002) (In Portuguese)
45. Fortes, M., Okos, M.R., Barret Jr, J.R.: Heat and mass transfer analysis of intra-kernel wheat drying and rewetting. J. Agric. Eng. Res. **26**, 109–125 (1981)
46. Fioreze, R.: The intermittent drying agricultural crops with particular reference to energy requirements. Doctorate thesis Cranfield Institute of Technology, University of Cranfield, UK (1986)

Grain Drying Simulation: Principles, Modeling and Applications

A. G. Barbosa de Lima, R. P. de Farias, S. R. Farias Neto, E. M. A. Pereira and J. V. da Silva

Abstract This chapter focuses in drying of wet solids in cross flow dryer (continuous drying system) with particular reference to grains. Here, topics related to grain, drying fundamental, types and selection of dryer, and drying models are presented in details. A cross flow dryer mathematical modeling that considers the influence of the porosity of the bed and transient terms in the drying process is presented and discussed. The governing conservation equations have been solved numerically using the finite-volume method and upwind formulation to convective terms. Application has been done to drying of yellow corn kernel. To analyze the influence of the main drying parameters on the quality of the product at the end of the process, results of the humidity ratio and temperature of the air and temperature and moisture content of the solid along of the drying process are presented, analyzed and compared with experimental data.

Keywords Drying · Finite-volume · Dryer · Cross-flow · Corn grain

A. G. Barbosa de Lima (✉) · E. M. A. Pereira · J. V. da Silva
Department of Mechanical Engineering, Federal University of Campina Grande, Av. Aprígio Veloso, 882, Bodocongó, Campina Grande, PB 58429-900, Brazil
e-mail: gilson@dem.ufcg.edu.br

E. M. A. Pereira
e-mail: evaldomarcos@bol.com.br

J. V. da Silva
e-mail: jvieira7@gmail.com

R. P. de Farias
Department of Agriculture Science, State University of Paraiba, Catolé do Rocha, PB, Brazil
e-mail: rp.eng.tec@gmail.com

S. R. Farias Neto
Department of Chemical Engineering, Federal University of Campina Grande, Av. Aprígio Veloso, 882, Bodocongó, Campina Grande, PB 58429-900, Brazil
e-mail: fariasn@deq.ufcg.edu.br

J. M. P. Q. Delgado and A. G. Barbosa de Lima (eds.), *Transport Phenomena and Drying of Solids and Particulate Materials*, Advanced Structured Materials 48,
DOI: 10.1007/978-3-319-04054-7_3,

1 Introduction

Many materials, for example, grain, fruits, vegetables, wood and clay products are considered to be hygroscopic capillary-porous products in which the pores are partially filled with liquid water and partially filled with an air/water-vapor mixture.

Grains, also called cereal or cereal grains are the seeds of grasses (*gramineae*). They are very rich in nutrients such as protein, vitamins and minerals. Grains are consumed in worldwide being important source of energy, carbohydrate, protein and fiber for the peoples, occupying the base of food pyramids. Examples of grains are corn, wheat, maize, rice, barley oats, and sorghum.

Grains are normally harvested still fairly moist, therefore, they must be dried to minimize the water content and to reduce spoilage problems due to the action of the microorganisms, in order for safe storage (grain must be stored for a period of time before it can be marketed or used as feed or seed) [1, 2]. Grain moisture content and temperature are important factors in determining safe storage life.

Drying is one the most widely known technique used to removal moisture in wet porous solids. Fast grain drying until storage moisture can produces higher moisture gradient into the grain. As this gradient reclines after drying, the grain surface receives moisture from the interior and expands while the grain interior loses moisture and contracts. The combination of tension and compression stresses (compressive at the surface and tensile at the centre of the grain) changes with time, thus, the grain tend to fail in tension by pulling itself apart at its centre.

Moisture absorption by low-moisture grains in the field before harvest is the most prevalent cause of fissured grains. Besides, most breakage in grain processing can also be attributed to grains which were fissured before milling. Mechanical harvesting of grains at higher moisture content can minimize this type of grain failure. Further, fissuring of the grain after drying can be controlled by the drying procedure.

The drying behavior of an individual particle, or of a collection of particles in a dryer, depends on the geometrical and physical characteristics of the solid. We notice that, in these processes the particles will vary in shape and properties, as a consequence also vary in drying behavior (moisture migration, heating, shrinkage and dilation rates). Then, because of the great applications in many industrial processes and scientific interest, to study the drying process plays an important role.

The selection of a particular type of dryer for a particular drying operation is a complex problem in virtue of the existence of many factors that affect the choice. In particular, factors to be considered during the drying process, which can be cited are, the drying techniques, heating method and the energy exchange between the product and the drying air. In general, the agricultural products (with particular reference to grains and seeds) are dried through two techniques: fixed bed (batch dryer) and continuous flow bed (concurrent, counter and cross-flow dryers). In the continuous flow dryers large quantities of grain can be dried without stopping, however, they present some disadvantages such as: initial high cost (normally including grain handling), need to be careful management (to obtain acceptable uniform drying), and need of matching loading and unloading equipment.

One of the most common continuous flow dryers for grains is the cross-flow dryer. In present day, because of the high cost of the dryers, in particular cross flow dryer and which produce considerable stress and cracking on the grain kernel, development of the new technologies for the drying sector, improvement of the product and rationalization of energy consumption, became necessary to devote study these driers with more details. In general, the drying residence time of grain in the drier has been obtained as a function of the following variables: layer thickness of the grain bed inside the dryer along airflow direction, initial and final moisture content of the grain, thermo-physical properties of the grains, air and grain flow rates, air and grain inlet temperature, and humidity ratio of the air.

In the drying area, physical, technical and economical problems associated to design, build and produce different dryers are important, very complicated and must be minimized. In present day, drying process mathematical modeling has the potential to quickly help in the identification of schedule alterations (and its consequences), to contribute significantly to the understanding of the moisture removal and heating mechanisms and others phenomena associated to the drying process, and solution of related problems.

In the last years, simulation has been an important tool for design and operation control. It provide a means for researcher, designer and engineer to elucidate the physics related to simultaneous heat and mass transfer phenomena which are generated within the wet porous solid (or porous media) during drying process, to propose control strategies, in order to find optimum design and operating parameters, to analyze the effect of disturbances, and to avoid expensive and repetitive experimentation. In this context, the computer-aided design systems for correct dryer design play an enormously important role [3–7]. Thus, a large number of the researchers have reported cross-flow dryer modeling. The models consider the void fraction (bed porosity) and/or the transient air-drying condition within the bed [1, 8–15]. Other researchers present numerical study considering void fraction and/or the transient terms in the mathematical model [16–22]. Experimental study has been reported in the literature too [21–28].

In order to obtain the better drying conditions and to save energy, it is necessary to know the effect of the drying parameters in the moisture removal and temperature of the solid during the drying process. In connection, now this chapter discusses mathematical modeling applied to drying process, in order to predict the changes of temperature and relative humidity of the air, and temperature and moisture content of the solid during drying process, with emphasis to continuous cross flow belt conveyor dryer.

2 Mathematical Modeling of Drying Process

Since past century that experimental and theoretical studies and analysis of the wet porous solids drying has been subject of many investigations. There is different forms to study theoretically wet solid drying process: (a) single particle or thin-layer of particles (particle model), and (b) deep-layer of particle (dryer model). Following we present discussion about this topic.

2.1 Thin-Layer Drying Model

For study drying of single particle (individual particle) two approaches has been cited in the literature: Distributed and lumped approaches. Distributed system analysis is the more actual method to study diffusion and convection transient problems. This method is used when internal gradients of the potential (for example moisture content and temperature) into the solid are significant. For instance, in drying, when significant temperature and moisture gradients occur inside the solid in every time, thus, moisture and temperature are functions of the position inside the solid and time, so, the distributed system analysis can be applied. However, when internal resistance to heat and mass fluxes is very small, heat and mass transfer takes place at the surface of the solid and uniform temperature and moisture distributions through the porous body is verified. Under this condition we can applied a lumped system analysis. This analysis is a simplest approach in diffusion and convection transient problems which may be used whenever diffusion inside the material is much faster than the diffusion across the boundary of the solid. In grain drying this approach is so called thin-layer drying model.

Mass diffusion is a phenomenon by which matter is transported from one region in space to another due to molecular motions. Thus, the diffusion theory states that diffusional flux is proportional to the concentration gradient (Fick's first law) [29–31]. The Fick's second law states that accumulated matter rates into the capillary-porous material equal to net mass transport by diffusion (effects of mass convection and mass generation aren't significant).

Lewis [32], Sherwood [33–35], Newman [36] and many other researchers (after Fick's second law of diffusion) states that the only moisture migration mechanism in capillary-porous bodies is the liquid diffusion [37]. Evaporation can take place at the surface of them. Thus, the Fick's second law has been frequently used to describe mass transfer inside wet porous bodies during drying process. The Fick's second law is given by:

$$\frac{\partial(\rho M)}{\partial t} = \nabla \cdot (\rho D \nabla M) \quad (1)$$

where ρ represent the density, M is the moisture content, D represent the mass diffusion coefficient, and t is the time. When the dry solid density is used in Eq. (1), the unit of the moisture content is kg of water per kg of dry matter, called dry basis. On the other hand, if wet solid density is used, so the unit of M is kg of water per kg of wet matter (kg of dry matter plus kg of water), called wet basis.

The average moisture content can be calculated as follows:

$$\bar{M} = \frac{1}{V}\int_{v} MdV \tag{2}$$

When we are using the Fick's second law of diffusion to describe drying process, truly we are utilizing a distributed approach.

Several semi-empirical and empirical models (thin-layer drying models) have been proposed to describe drying process take into account only external resistance to heat and moisture transfer between grain and air (internal resistance is negligible). In this analysis, lumped parameters typically combine aspects of several different physical phenomena such as shrinkage, heterogeneity and shape of the body, and under appropriate situation may actually give a better result than a more complex formulation, for example, a distributed model. Besides require less input data to theoretically study the physical problem.

In the thin-layer model, average moisture content can be given as follows:

$$\overline{M} = f(T,\ RH, w,\ M_0, t,\ S,\ V,\ etc.) \tag{3}$$

where T, RH and w represent temperature, relative humidity and velocity of the air, respectively, and M_0, V and S represent the initial moisture content, volume and surface area of the solid. More details about this topic can be cited in the references cited in the text. The practical importance of thin-layer drying is very limited, because the materials are usually dried in deep-layers: stationary or moving. Further, unfortunately no theory has been developed that accurately and practically describes the thin-layer drying rate.

2.2 Deep-Bed Drying Model

The drying of a single particle individually or even of a layer of material of small thickness, does not modify the drying conditions significantly. However, when the material is placed inside of a dryer, forming a thick layer, the thermodynamic properties of the drying air are modified markedly.

In the drying of grain in a deep-bed, whilst individual grain may all is losing moisture at different rates. Thus, the air absorbs moisture (vapor) as it moves through the bed until it becomes effectively saturated and moves through the remaining layers of grain without effecting further drying. In this case, their drying potential decreases. In virtue of this, the models of drying in thick layer are more complete and more complex than the thin layer.

The dryer models used by researchers take into account thermo-physical properties, drying kinetics and conservation of mass and energy in the dryer. It, therefore, satisfies the need to have an equation for drying of the material in a thin layer under certain pre-established operational conditions.

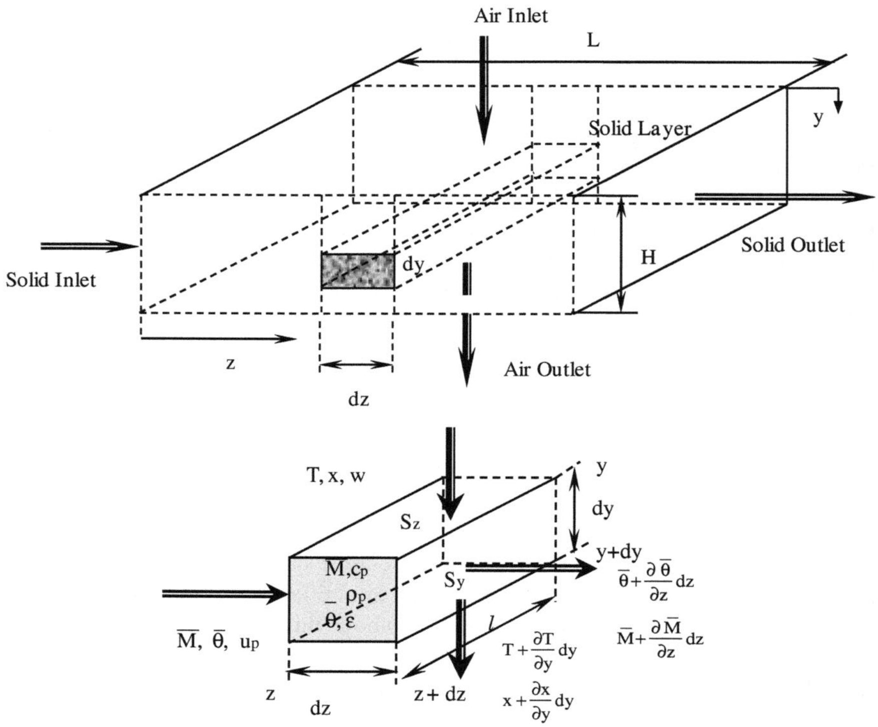

Fig. 1 Schematic representation of the continuous cross flow dryer

In general, the deep-layer drying mathematical model consists of four hyperbolic partial differential equations with four unknowns [38] and can really predict the performance of the dryer.

In multi-zone cross-flow dryer (Fig. 1), air flows in the y direction and solid particles in the z direction. The length of the drier and the number of drying zone are fixed by process conditions. For this dryer type, the transient model presented herein is based in the Michigan State University model reported by Bakker-Arkema et al. [8]. The development of the conservation equation is based on the control-volume illustrated in Fig. 1 [27, 28]. As a simplification of the model to describe the drying process of solids in a cross-flow dryer, the following assumptions have been made:

(a) The volume variation is negligible during the drying process, i.e., the solid is considered non-deformable;
(b) The temperature and moisture content gradients within the individual particle are negligible along the process.
(c) Heat conduction among the particles is negligible.
(d) Heat loss of the dryer to the surrounding is negligible.

(e) Air and grain flows are plug-type.
(f) The moisture evaporation takes place at the drying-air temperature.
(g) The inter-particle void fraction (bed porosity) is constant.

According to assumptions, the following partial differential equations are obtained to model the transient cross flow drying process in the dryer:

2.2.1 Mass Transfer Models

From a mass balance of the air into the infinitesimal volume presented in Fig. 1, we have that the amount of water vapor that enters the area S_z at position y, minus the amount of water vapor which leaves the area S at position y + dy, plus the change in air humidity in the open spaces is equal to the moisture provided by the product. Thus, we can write:

$$\frac{\partial(\rho_a x)}{\partial t} + \frac{\partial}{\partial y}\left(\rho_a \frac{w_a}{\varepsilon} x\right) = -\frac{\rho_P}{\varepsilon}\frac{\partial \overline{M}}{\partial t} \quad (4)$$

For the solid we notice that the moisture content of the product can be obtained by using an empirical equation of thin layer, suitable for each product.

$$\frac{\partial \overline{M}}{\partial t} = Thin\text{-}layer\, drying\, equation \quad (5)$$

In the Eq. (4), ρ_a is the air density, x is the air humidity ratio, w_a is the air velocity, ε is the bed porosity (void volume per volume of the dryer), ρ_p is the product density, and y is the Cartesian coordinate.

2.2.2 Energy Transfer Models

From the energy balance of the air into the infinitesimal volume pictured in Fig. 1, we have that the energy that enters the area S_z in the y position, minus the energy leaving the area S_z at position y + dy, is equal to the energy transferred to the product, by convection, and the variation of the energy with respect to time of the air in the empty spaces. Thus,

$$\frac{\partial}{\partial t}(\rho_a T) + \frac{\partial}{\partial y}\left(\rho_a \frac{w_a}{\varepsilon} T\right) = -\frac{A^* h_c(T - \overline{\theta})}{\varepsilon(c_a + x c_v)} \quad (6)$$

where T is the air temperature, A* is the specific surface area (surface area of the solid per volume unit of the bed), h_c is the convective heat transfer coefficient, $\overline{\theta}$ is the average temperature of the solid, c_a and c_v are the specific heat of the air and vapor, respectively.

From a energy balance in the solid, we have that the energy transferred by convection from air to the product, is equal to the energy required to heat the product, plus the energy required to evaporate water from the product, and the energy required to heat the evaporated water vapor. Then, we have,

$$\frac{\partial}{\partial t}(\rho_p \bar{\theta}) = \frac{A^* h_c (T - \overline{\theta})}{c_p + c_w \bar{M}} + \frac{\left[h_{fg} + c_v (T - \overline{\theta})\right]}{c_p + c_w \bar{M}} \rho_p \frac{\partial \overline{M}}{\partial t} \quad (7)$$

where h_{fg} is the heat of vaporization of water into the product and c_w is the specific heat of water.

For a well-posed transient mathematical formulation, we necessitate specify the initial and boundary conditions. The following one were used:

$$\overline{M}(y, t = 0) = M_0 \quad (8)$$

$$\overline{\theta}(y, t = 0) = \theta_0 \quad (9)$$

$$\mathrm{T}(\mathrm{y} = 0, \mathrm{t}) = \mathrm{T}_0; \quad (10)$$

$$\mathrm{x}(\mathrm{y} = 0,\ \mathrm{t}) = \mathrm{x}_0 \quad (11)$$

3 Application: Corn Grain Drying in Cross-Flow Dryer

3.1 Physical Problem

Knowledge of the structure, composition, and property values is useful in understanding the drying and storage characteristics of grains. The kernel structure may affect the drying rate. The chemical composition of corn grain at 14 % (wet basis) moisture content is 9.8 % protein, 5 % fat, 63.6 % starch, and 2.0 % fiber. The principal parts of a corn are the pericarp (5.0–7.0 % weight), germ (10.0–12.0 % weight), and the endosperm (82.0–84.0 % weight) [1].

Corn grain is usually harvested at high moisture content for safe storage. Since it is hygroscopic will lose or gain moisture until equilibrium is reached with the surrounding air. So, it is necessary to dry this agricultural product to storage and to prevent quality deterioration. As application, the methodology was used to describe drying of yellow corn grains. In this sense, Brokker et al. [1] report the following thin-layer drying equation to describe drying rate:

$$\frac{\partial \overline{M}}{\partial t} = \frac{M_e - \overline{M}}{3600\left[A^2 + \frac{Bt}{900}\right]^{1/2}} \quad (12)$$

where t is in seconds, M_e is the equilibrium moisture content, $A = -1.7054824 + 0.0087917\,\overline{\theta}$ and $B = 148.60862 \times 10^{-0.059418\overline{\theta}}$.

The heat of vaporization, equilibrium moisture content, specific surface area, dry solid density and specific heat of the corn grain, and void fraction of the bed are given by [1], as follows:

$$h_{fg} = (2502.2 - 2.39T)\left[1 + 1.2925 \times 10^{-16.981\overline{M}}\right] \quad kJ/kg \tag{13}$$

$$M_e = \frac{1}{100}\left[-\frac{\ln(1-x)}{8.6541 \times 10^{-5}(T + 49.81)}\right]^{1/1.8634} \tag{14}$$

$$c_p = \left[1.361 + 3.97\frac{\overline{M}}{(1+\overline{M})}\right] \quad kJ/(kgK) \tag{15}$$

with $\rho_P = 650$ kg/m^3, $\varepsilon = 0.44$ and $A^* = 784\,m^2/m^3$.

The specific heat [39], density, relative humidity, absolute temperature, universal constant applied to the air, saturation pressure of vapor and local atmospheric pressure are given by [40]:

$$c_a = 1.00926 - 44.04033 \times 10^{-5}T + 6.17596 \times 10^{-7}T^2 - 4.0972 \times 10^{-10}T^3 kJ/(kgK) \tag{16}$$

$$\rho_a = \frac{P_{atm}M_a}{RT_{abs}} \tag{17}$$

$$RH = \frac{P_{atm}x_a}{(x_a + 0.622).P_{vs}} \tag{18}$$

$$T_{abs} = T + 273.15\ K \tag{19}$$

with $R = 8314.34$ J/(kgK), $P_{atm} = 101325$ Pa and $M_a = 28.97$ kg/kmol,

$$P_{vs} = 22105649.25Exp\{[-27405.53 + 97.5413T_{abs} - 0.146244T_{abs}^2 + 0.12558 \times 10^{-3}T_{abs}^3 - 0.48502 \times 10^{-7}T_{abs}^4]/[4.34903T_{abs} - 0.39381 \times 10^{-2}T_{abs}]\} \quad Pa \tag{20}$$

The specific heat of water on the vapor and liquid phases are given by [39]:

$$c_v = 1.8830 - 0.16737 \times 10^{-3}T_{abs} + 0.84386 \times 10^{-6}T_{abs}^2 - 0.26966 \times 10^{-9}\,T_{abs}^3\,kJ/(kgK) \tag{21}$$

$$c_w = 2.82232 + 1.18277 \times 10^{-2}\,T_{abs} - 3.5047 \times 10^{-5}\,T_{abs}^2 + 3.6010 \times 10^{-8}T_{abs}^3 kJ/(kgK) \tag{22}$$

The heat transfer coefficient was obtained using the following equations [1]:

$$h_c = \begin{cases} 101.4(\rho_a w_a)^{0.59} \text{ to } \rho_a w_a \geq 0.68 \\ 99.6(\rho_a w_a)^{0.49} \text{ to } \rho_a w_a < 0.68 \end{cases} \mathrm{W/(m^2\,{}^{o}C)} \tag{23}$$

3.2 Numerical Treatment

3.2.1 Discretization of the Governing Equations

Many numerical techniques can be used to solve the set of partial differential equations, for example, finite-element, finite-difference, boundary-element and finite-volume methods [41–44]. In this work, the finite-volume method was used to discretize the basic equations by integrating one under the control volume and time as illustrated in Fig. 2. In this figure P represent the central nodal point and S and N are the neighbor nodal points.

The integrations of the model equations will results in different set of linear equations in the discretized form. For all equations was used upwind scheme as interpolation function to convective terms along the z-direction. Thus, for the grain we have the following discretized energy transfer equation:

$$A_P \overline{\theta}_P = A_P^o \overline{\theta}_P^o + S_C^{\overline{\theta}} \tag{24}$$

where

$$A_p = \frac{\Delta z}{\Delta t_m} + \frac{h_c\, A^* \Delta z}{\rho_p\, c_p\, + \rho_p\, c_w\, \overline{M}} + \frac{c_v \rho_p\, \frac{\partial \overline{M}}{\partial t} \Delta z}{\rho_p\, c_p\, + \rho_p\, c_w\, \overline{M}} \tag{25}$$

$$A_P^o = \frac{\Delta z}{\Delta t_m} \tag{26}$$

$$S_C^{\overline{\theta}} = \frac{(h_{fg} + c_v T_p)\, \rho_p\, \frac{\partial \overline{M}}{\partial t}}{\rho_p\, c_p\, + \rho_p\, c_w\, \overline{M}} \Delta z + \frac{h_c\, A^* T_p \Delta z}{\rho_p\, c_p\, + \rho_p\, c_w\, \overline{M}} \tag{27}$$

and the following discretized mass transfer equation:

$$A_P \overline{M}_P = A_P^o \overline{M}_P^o + S_C^{\overline{M}} \tag{28}$$

Fig. 2 Control-volume used in this work

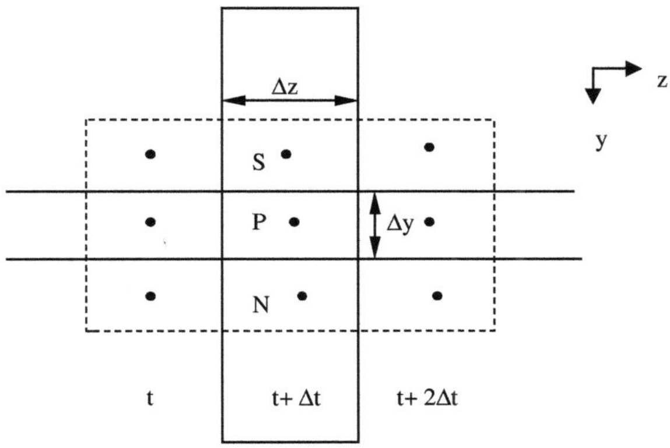

where

$$A_P = \frac{\Delta z}{\Delta t_m} + \frac{\Delta z}{3600(A^2 + \frac{Bt}{900})^{1/2}} \tag{29}$$

$$A_P^o = \frac{\Delta z}{\Delta t_m} \tag{30}$$

$$S_C^{\bar{M}} = \frac{M_e \Delta z}{3600(A^2 + \frac{Bt}{900})^{1/2}} \tag{31}$$

For the air we have the following discretized energy transfer equation:

$$\mathrm{A_P T_P = A_S T_S + A_P^o T_P^o + S_C^T} \tag{32}$$

where

$$A_P = \frac{\Delta y}{\Delta t} + \frac{w_a}{\varepsilon} + \frac{A * h_c \Delta y}{\varepsilon(\rho_a c_a + \rho_a x c_v)} \tag{33}$$

$$A_S = \frac{w_a}{\varepsilon} \tag{34}$$

$$A_P^o = \frac{A * h_c \Delta y}{\varepsilon(\rho_a c_a + \rho_a x c_v)} \tag{35}$$

$$S_C^T = \frac{\Delta y}{\Delta t} \tag{36}$$

and the following discretized mass transfer equation:

$$A_P x_P = A_S x_S + A_P^o x_P^o + S_C^x \tag{37}$$

where

$$A_P = \rho_a \frac{\Delta y}{\Delta t} + \rho_a \frac{w_a}{\varepsilon} \tag{38}$$

$$A_S = \rho_a \frac{w_a}{\varepsilon} \tag{39}$$

$$A_P^o = \rho_a \frac{\Delta y}{\Delta t} \tag{40}$$

$$S_C^x = -\frac{\rho_p}{\varepsilon} \frac{\partial \overline{M}}{\partial t} dy \tag{41}$$

To obtain the numerical results, a computational code using the software Mathematica® was implemented. In the equations applied to the air, the time step was evaluated by $\Delta t = \Delta y / w_a$. For the grains, $\Delta z = u_p(npy - 1)\Delta t$ and $\Delta t_m = (npy - 1)\Delta t$, where npy is the nodal point number in the direction y, and u_p represent the product velocity into the dryer. During this time step Δt_m, the grains within the volume $\Delta z H$ were assumed stationary and thus $T, x, \overline{M}$ and $\overline{\theta}$ were obtained as for the fixed bed drying.

3.2.2 Numerical Procedure for Condensation of Water

After calculates $\overline{M}$, $\overline{\theta}$, T and x in each position inside bed and drying time, the relative humidity is obtained. If this value is greater than 1, then saturation or super saturation is assumed and condensation is modeled. The condensation of water may occur when a large amount of vapor of water is transported by the air and so cooled as it passes through cool grains

To model the condensation the following procedure was used:

(a) In a y location inside the bed are calculated $\overline{M}$, $\overline{\theta}$, Tex, and so is obtained P_{vs} and RH.
(b) If RH > 1, then is assumed a new value of the absolute humidity $x = x - \Delta x$. So, go to passes c. If RH $\leq$ 1 stop the condensation and go to new y location.
(c) Using the new value x, we calculate the new values of $\overline{M}$, $\overline{\theta}$ and T.
(d) Using the new value of T, we calculate P_{vs} and RH and return to passes b.

In this formulation, the new value of average moisture content $\overline{M}$ is given by:

$$\overline{M} = \overline{M}_{bc} + \left(\frac{\rho_a W_a \Delta z}{\rho_p u_p \Delta y}\right)(x_{bc} - x) \tag{42}$$

The new value of air temperature T is given by:

$$T = \frac{\frac{\rho_a w_a \Delta z}{u_p}(c_a + c_v x_{bc})T_{bc} + \rho_p \Delta y (c_p + c_w \overline{M}_{bc})\overline{\theta}_{bc}}{\rho_a w_a \frac{dz}{u_p}(c_a + c_v x)} + \frac{-\rho_p \Delta y (c_p + c_w \overline{M}_{bc})\overline{\theta}_{bc} - \rho_a w_a (\frac{\Delta z}{u_p}) h_{fg}(x - x_{bc})}{\rho_a w_a \frac{\Delta z}{u_p}(c_a + c_v x)} \quad (43)$$

In this procedure was used $\Delta x = 10^{-8}$ kg water/kg dry air. The subscript "bc" refers to values before condensation.

3.3 Results and Analysis

In order to analyze the effects of the drying-air conditions on the heating rate and moisture removal of the yellow corn grain, several conditions were chosen for simulation. Table 1 presents all drying condition used in the work as well as the final moisture content, total drying time and length of the dryer.

To validate the methodology, numerical results of the average moisture content of yellow corn grain are compared with experimental data to fixed bed drying reported in the literature [45]. The comparison is possible because $u_p << w_a$ ($u_p = 0.005$ m/s). Figure 3 illustrates this comparison during drying process in $y \approx 0.0$ m (test 1). It is verified that a lower deviation between the results was obtained.

Figures 4 and 5 show the effect of inlet air-drying conditions on the moisture removal and temperature of the grain. It is verified that the drying temperature has strong effect on the grain temperature more than on the moisture content. However, the increase of the air temperature increases the drying rate and the grain reached the temperature of the air and its equilibrium moisture content more quickly. This situation provokes higher moisture and thermal gradients and higher hydro and thermal stresses within the grain, which may cause damage to the grain and reduces its quality.

Figures 6 and 7 show both air and grain temperatures within the bed at different instants and conditions of the drying-air. It is verified that the highest gradients occurs in the grain in few instants of drying and close to entrance of the air in the dryer.

The high thermal gradients along the bed are not recommended because it produces non uniform drying and big thermal stress in the grain and may cause cracking, fissures and deformation in the solid and to reduce its quality at the end of the process.

It can be observed in the Fig. 6b that air temperature only arrive until a certain limit, due to the relative humidity inside the bed to reach its maximum value at few instants of drying. In this same drying period the grain temperature (Fig. 7b)

Table 1 Air and grain condition used in the simulations, total drying time and length of the dryer

Test	Grain					Air			t (s)	L (m)
	M_0 (kg/kg)	H (m)	θ_0 (°C)	M_f^{**} (kg/kg)	x_0 (kg/kg)	w_a (m/s)	T (°C)	RH (%)		
1	0.30	0.1	24	0.180	0.011340700	1.63	75	4.7	4907.90	24.54
2	0.30	0.2	25	0.231	0.002628163	1.50	30	10.0	8000.00	40.00
3	0.30	0.2	25	0.215	0.004583815	1.50	40	10.0	8000.00	40.00
4	0.30	0.2	25	0.177	0.012514130	1.50	60	10.0	8000.00	40.00
5	0.30	0.2	25	0.158	0.019776720	1.50	70	10.0	8000.00	40.00
6	0.30	0.2	25	0.194	0.000761409	1.50	50	1.00	8000.00	40.00
7	0.30	0.2	25	0.195	0.003825780	1.50	50	5.00	8000.00	40.00
8	0.30	0.2	25	0.196	0.007698940	1.50	50	10.0	8000.00	40.00
9	0.30	0.2	25	0.198	0.015590900	1.50	50	20.0	8000.00	40.00
10	0.30	0.2	25	0.200	0.031983900	1.50	50	40.0	8000.00	40.00
11	0.30	0.2	25	0.202	0.049242300	1.50	50	60.0	8000.00	40.00
12	0.30	0.2	25	0.134	0.002923429	1.50	80	1.00	8000.00	40.00
13	0.30	0.2	25	0.141	0.030526000	1.50	80	10.0	8000.00	40.00
14	0.30	0.2	25	0.146	0.064204200	1.50	80	20.0	8000.00	40.00
15	0.30	0.2	25	0.154	0.143195100	1.50	80	40.0	8000.00	40.00
16	0.30	0.2	25	0.163	0.242745300	1.50	80	60.0	8000.00	40.00
17	0.30	0.1	25	0.188	0.014897320	1.50	80	5.00	4000.00	20.00
18	0.30	0.2	25	0.138	0.014897320	1.50	80	5.00	8000.00	40.00
19	0.30	0.3	25	0.138	0.014897320	1.50	80	5.00	8000.00	40.00
20	0.30	0.5	25	0.152	0.014897320	1.50	80	5.00	6666.67	33.33
21	0.30	1.0	25	0.138	0.014897320	1.50	80	5.00	8000.00	40.00
22	0.30	0.1	25	0.226	0.003825788	1.50	50	5.00	4000.00	20.00
23	0.30	0.3	25	0.195	0.003825788	1.50	50	5.00	8000.00	40.00
24	0.30	0.5	25	0.204	0.003825788	1.50	50	5.00	6666.67	33.33
25	0.30	1.0	25	0.195	0.003825788	1.50	50	5.00	8000.00	40.00
26	0.30	0.2	25	0.195	0.003825788	0.10	50	5.00	8000.00	40.00
27	0.30	0.2	25	0.195	0.003825788	0.30	50	5.00	8000.00	40.00
28	0.30	0.2	25	0.195	0.003825788	0.50	50	5.00	8000.00	40.00
29	0.30	0.2	25	0.195	0.003825788	1.00	50	5.00	8000.00	40.00

**In y = 0.0 m

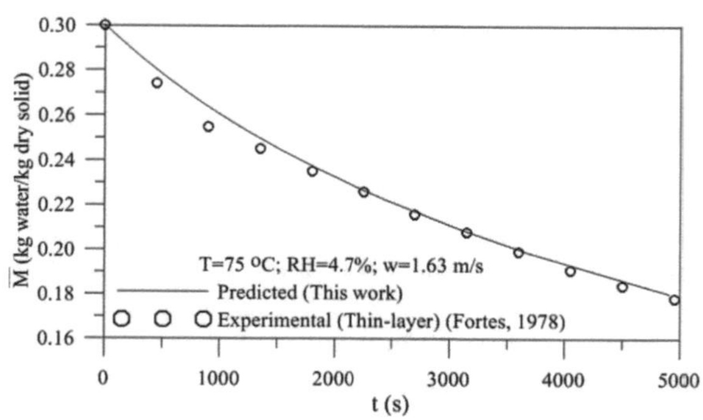

Fig. 3 Comparison between the predicted and experimental values [45] of the average moisture content in y ≈ 0.0 m during drying process of yellow corn grain (test 1)

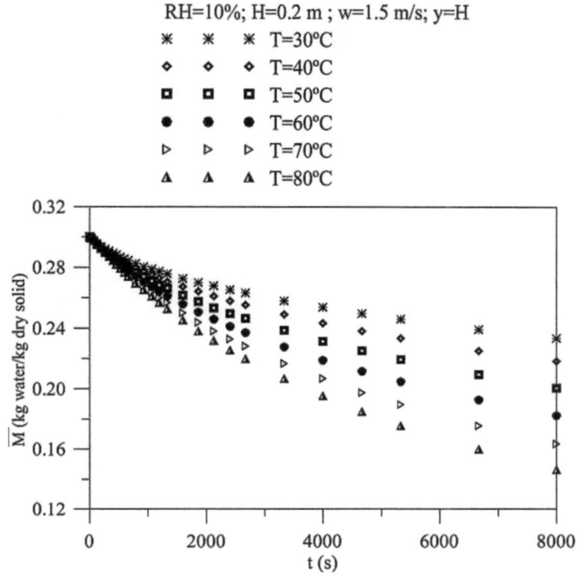

Fig. 4 The effect of the air-drying temperature under the average moisture content of the grain during drying process

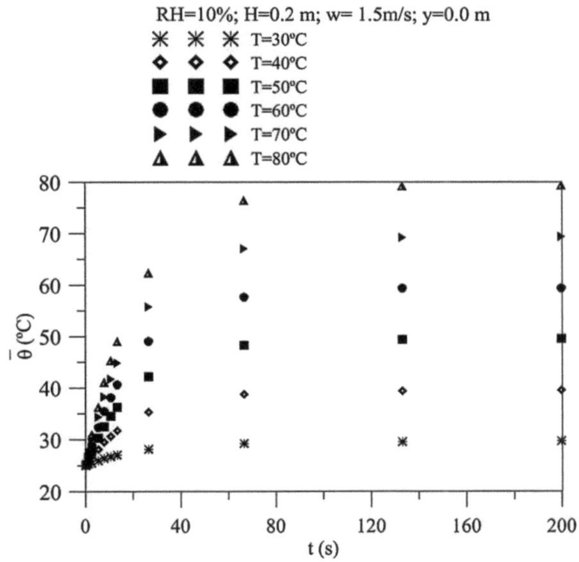

Fig. 5 The effect of the air-drying temperature under the grain temperature during drying process

presents small thermal gradients in both 1 and 10 % relative humidity. The high relative humidity does with that the grain temperature reaches temperature near the wet bulb temperature of air quickly, being stabilized soon after. It is noticed that at the surface of the product, the temperature of the corn equals to the wet bulb temperature of the air. In smaller drying times and in the first layers of products,

Fig. 6 Air temperature distribution within the bed to nine drying times. **a** RH = 1 % and **b** RH = 10 %

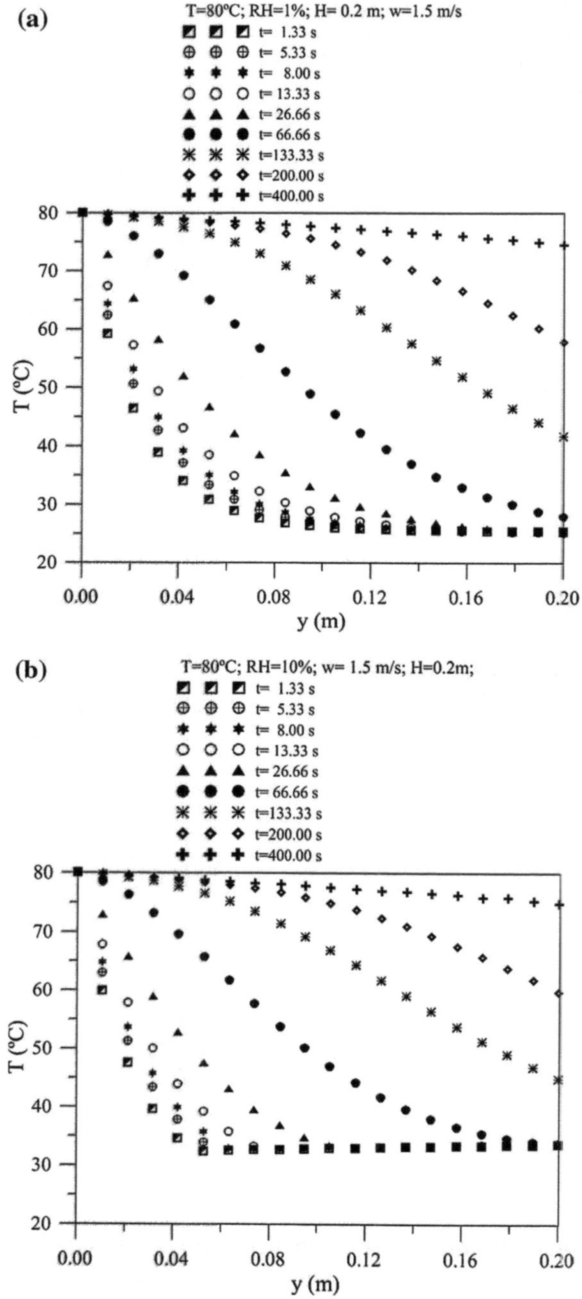

Fig. 7 Grain temperature distribution within the bed to nine drying times. **a** RH = 1 % and **b** RH = 10 %

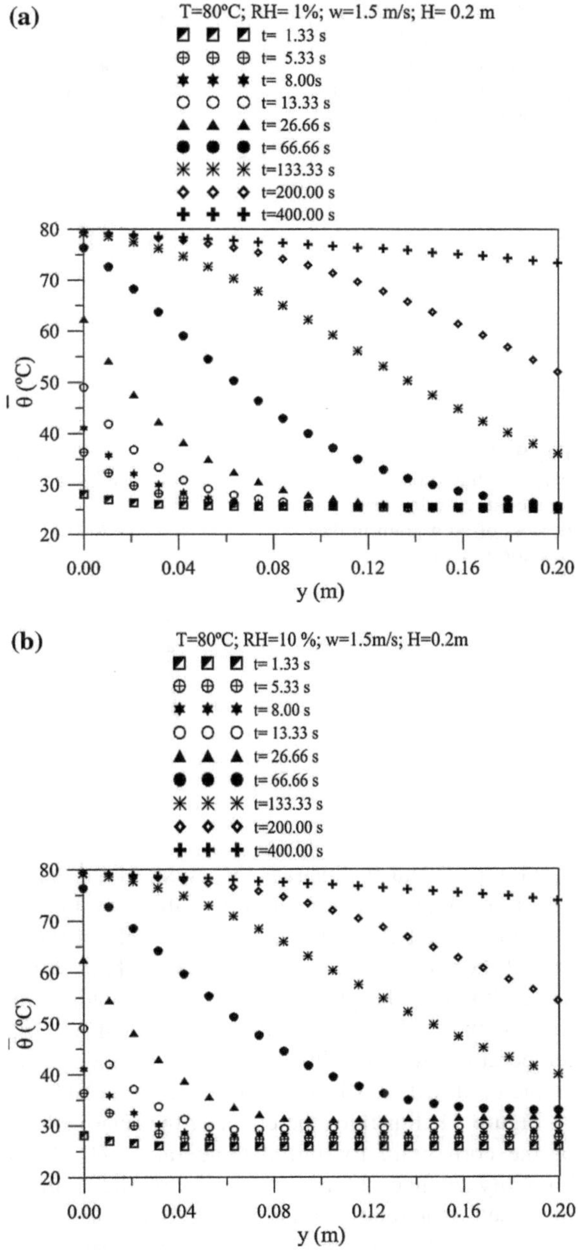

the grain temperature rises slightly, due to the heating of the air and later it stays constant for the other points. This happens because the air to reach relative humidity of 100 % in these positions.

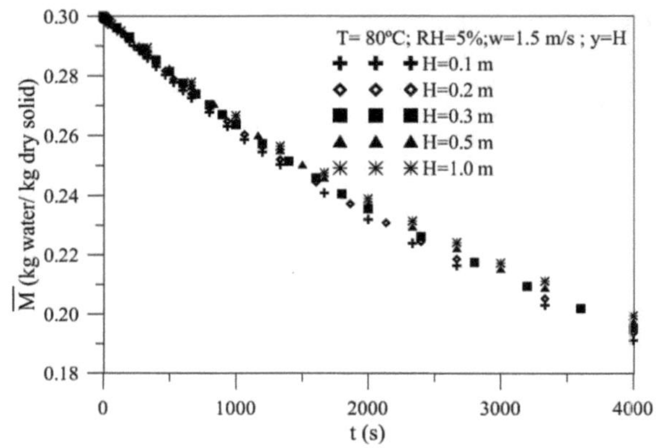

Fig. 8 The effect of layer height in the moisture removal during drying process of corn grain in the position y = H

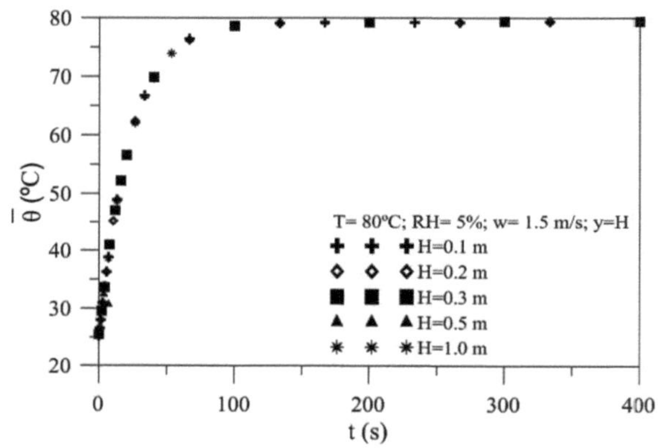

Fig. 9 The effect of layer height in the grain temperature during drying process of corn grain in the position y = H

Figures 8 and 9 illustrate the effect of the height of the product layer in the moisture content and temperature of the grain in the outlet of layer. In these figures it is observed that the increase in the height of the layer provides a smaller drying rate of the product, because the air is reaching saturation along of the layer. In the same situation, the product reaches the air temperature in few minutes after initialized the drying process as before. Then, this parameter affects not the grain temperature strongly. The same behavior is verified for different relative humidity as pictured in Fig. 10 in the first layer (y $\approx$ 0.0 m).

Figure 11 illustrates the effect of the relative humidity in the moisture removal of the corn grain for initial temperature 80 °C. It is verified that the relative humidity changes moisture content of the grain, for a fixed temperature. The increase on the relative humidity affect the drying rate as expected. Then, with the decrease of RH, moisture content of the corn grain decreases more quickly along the process, thus, reducing the air temperature in the bed. When saturation occur in the bed, the air temperature reaches the wet bulb temperature, producing high temperature gradient mainly in the first layers of the bed and for higher air relative humidity (RH > 40 %) at the inlet of the bed.

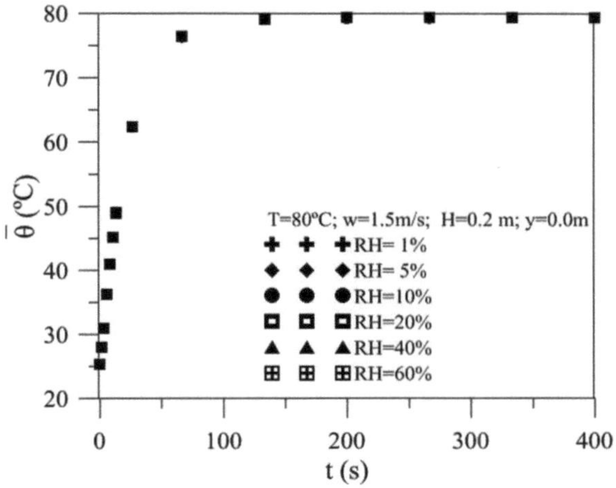

Fig. 10 The effect of relative humidity in the corn grain temperature during drying process (T = 80 °C)

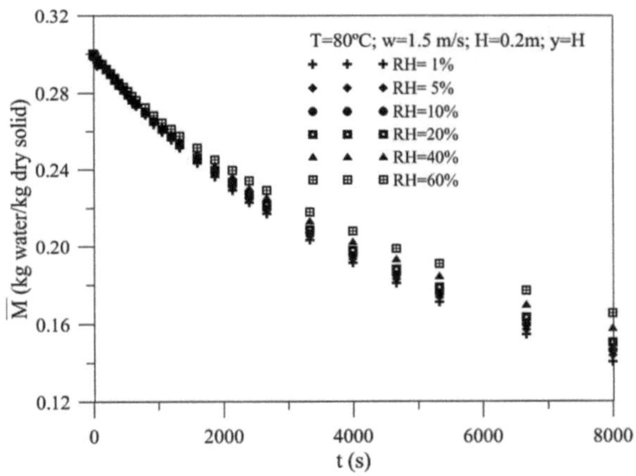

Fig. 11 The effect of relative humidity in the moisture removal during drying process of corn grain (T = 80 °C)

Figures 12 and 13 illustrate the effect of air velocity in the moisture removal and temperature of corn grain during drying process, respectively. An increase in the airflow rate caused a decrease of the moisture gradients and also an increase on the drying rate of solid. Considerable effects were verified on the heating rate of the grain. By increasing of the air velocity, the grain reaches the equilibrium temperature more quickly. Thus, the drying process is controlled by internal and external diffusion.

Because drying is a very energy consumption process, in recent years there has been a substantial development in reducing the energy consumption in driers. This development has been moving in two directions: improvement of the actual drying processes to make them consume less energy, and improvement of heat recovery systems. Therefore, to low thickness layer and small relative humidity, the air in the outlet of the dryer can be recirculate and used to dry solid and to save energy.

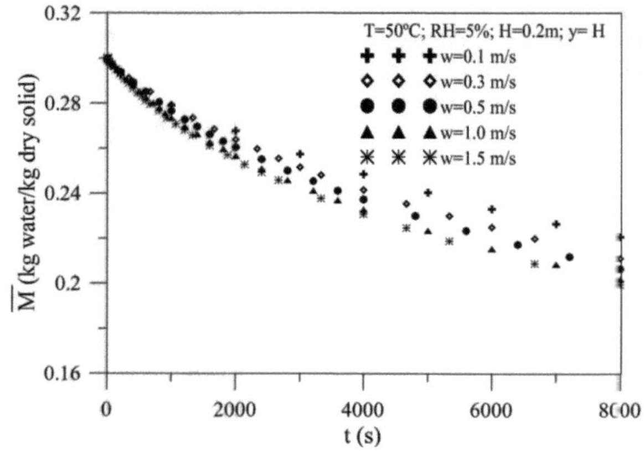

Fig. 12 The effect of air velocity in the moisture removal of corn grain during drying process (T = 50 °C)

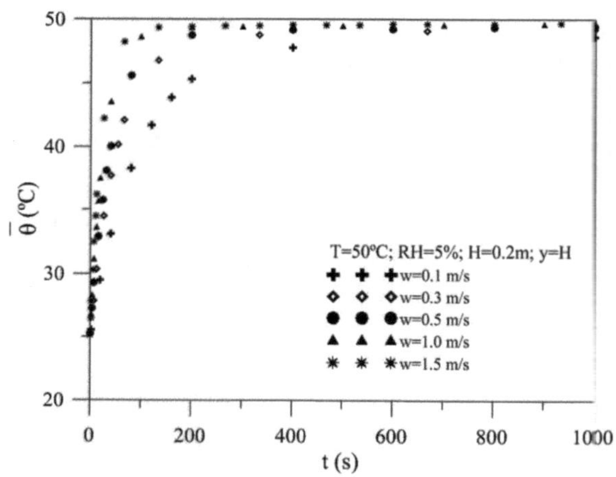

Fig. 13 The effect of air velocity in the corn grain temperature during drying process (T = 50 °C)

4 Concluding Remarks

This chapter focuses advanced topics related to heat and mass transfer in cross-flow dryer with particular reference to drying process. This type of physical problem is motivated by its importance in many industrial processes such as mineral processing, food processing, and chemical processes. Fundamentals information about principle of operation, design, selection, and interactions between the product and air inside the dryer are given. A transient mathematical modeling based on the mechanistic approach is introduced to explain the complex heat and mass transfer between the moist particle and gas phase inside the continuous cross-flow dryer.

Application to predict drying process of corn grain is performed and numerical studies were conducted for different inlet air conditions. From the analysis of the predicted it was verified that the finite-volume method can be used to simulate drying process in cross flow dryer because the good agreement obtained by comparison between numerical and experimental data of the average moisture content.

It is possible to conclude that the mass transfer is controlled by internal diffusion, and external condition, because airflow rate affect the drying rate. During drying process low moisture content gradients within the bed were obtained. This is due to the small thickness layer of the grain used in the simulation. For high relative humidity (around 100 %) occurred condensation of water in the bed. Thus, the air temperature equals to the wet bulb temperature of the air, mainly in the few instants of drying. Further, as both the thickness layer of the product (corn grain) and the relative humidity of the drying air increases, drying rate decreased, however, practically, it doesn't affect heating rate of the product.

The air temperature has effect on the drying rate of corn grain more than airflow rate. The higher air temperature gradients inside of the bed, happen in the few instants of drying (400.00 s), for any air relative humidity. Thus, the grain reaches the inlet air temperature in few minutes of drying for all drying conditions.

Acknowledgments The authors would like to express their thanks to CNPq (Conselho Nacional de Desenvolvimento Científico e Tecnológico, Brazil), CAPES (Coordenação de Aperfeiçoamento de Pessoal de Nível Superior, Brazil), and FINEP (Financiadora de Estudos e Projetos, Brazil) for supporting this work; to the authors of the references in this chapter that helped in our understanding of this complex subject, and to the Editors by the opportunity given to present our research in this book.

References

1. Brokker, D.B., Bakker-Arkena, F.W., Hall, C.W.: Drying and Storage of Grains and Oil Seeds. AV1 Book, New York (1992)
2. Rumsey, T.R., Rovedo, C.O.: Two-dimensional simulation model for dynamic cross-flow rice drying. Chem. Eng. Process. **40**, 355–362 (2001)
3. Pang, S., Haslett, A.N.: High-temperature kiln drying of softwood timber: the role of mathematical modeling. In: Turner, I., Mujumdar, A.S. (eds.) Mathematical Modeling and Numerical Techniques in Drying Technology. Marcel Dekker Inc, New York (1997)
4. Houška, K., Valchář, J., Viktorin, Z.: Computer-aided design of dryers. In: Mujumdar, A.S. (ed.) Advances in Drying, vol. 4. Hemisphere Publishing Corporation, New York (1987)
5. Sokhansanj, S.: Grain drying simulation with respect to energy conservation and grain quality. In: Mujumdar, A.S. (ed.) Advances in Drying, vol. 3. Hemisphere Publishing Corporation, New York (1984)
6. Imre, L.: Solar drying. In: Mujumdar, A.S. (ed.) Handbook of Industrial Drying, vol. 1. Marcel Dekker, Inc., New York (1995)
7. Turner, I., Perré, P.: A synopsis of the strategies and efficient resolution techniques used for modelling and numerically simulating the drying process. In: Turner, I., Mujumdar, A.S. (eds.) Mathematical Modeling and Numerical Techniques in Drying Technology. Marcel Dekker, Inc., New York (1997)
8. Bakker-Arkema, F.W., Lerew, L.E., De Boer, S.F., Roth, M.C.: Grain drying simulation, Reseacher Report 224 (1974)

9. Sokhansanj, S., Wood, H.C.: Simulation of thermal and disinfestations characteristics of forage dryer. Drying Technol. **9**(3), 643–656 (1991)
10. Fasina, O., Sokhansang, S.: Modeling the bulk cooling of alfalfa pellets. Drying Technol. **13**(889), 1881–1904 (1993)
11. Barrozo, M.A.S., Sartori, D.J.M., Freire, J.T.: Simultaneous heat and mass transfer during the drying of the soybeans seeds in a crossflow moving bed. In: Proceeding of the International Drying Symposium (IDS), Krakow, Poland, vol. B, pp. 873–880, (1996)
12. Motta-Lima, O.C., Pinto, J.C., Massarani, G.: Parameter estimation in cross-flow sliding bed drying. In: Proceedings of the International Drying Symposium, Krakow, Poland, vol. A, pp. 283–290, (1996)
13. Li, Y., Cao, C., Liu, D.: Simulation of recirculating circular grain dryer with tempering stage. Drying Technol. **15**(1), 201–214 (1997)
14. Liu, Q., Bakker-Arkema, F.W.: Automatic control of cross-flow dryers, part 2: design of a model-predictive controller. J. Agric. Eng. Res. **80**(2), 173–181 (2001)
15. Liu, Q., Bakker-Arkema, F.W.: Automatic control of cross-flow dryers, part 1: development of process model. J. Agric. Eng. Res. **80**(1), 81–86 (2001)
16. Eltigani, A.Y., Bakker-Arkena, F.W.: Automatic control of commercial cross-flow grain dryers. Drying Technol. **5**(4), 561–575 (1987)
17. Vasconcelos, L.G.S., Alsina, O.L.S.: Drying simulation of (carioca(beans in cross-flow. In: Proceedings of the International Drying Symposium (IDS), Montreal, Canada, vol. B, pp. 1500–1507, (1992)
18. França, A.S., Fortes, M., Haghighi, K.: Numerical simulation of intermittent and continuous deep-bed drying of biological material. Drying Technol. **12**(7), 1537–1560 (1994)
19. Soponronnarit, S., Prachaayawarakorn, S., Sripawatakul, O.: Developments of cross-flow fluidized bed paddy dryer. Drying Technol. **14**(10), 2397–2410 (1996)
20. Giner, S.A., Mascheroni, R.H., Wellist, M.E.: Cross-flow drying of wheat: a simulation program with a diffusion-based deep-bed model and a kinetic equation for viability loss estimation. Drying Technol. **14**(10), 2255–2292 (1996)
21. Giner, S.A., Bruce, D.M., Mortimore, S.: Two-dimensional model of steady-state mixed-flow grain drying: part 1: the model. J. Agric. Eng. Res. **71**, 37–50 (1998)
22. Giner, S.A., Bruce, D.M.: Two-dimensional model of steady-state mixed-flow grain drying: part 2: experimental validation. J. Agric. Eng. Res. **71**, 51–56 (1998)
23. Santori, D.J.M.: Drying of reeds in cross-flow moving bed: mechanical effect and air humidity. In: Proceedings of the International Drying Symposium (IDS), Montreal, Canada, vol. B, pp. 1524–1533, (1992)
24. Pimentel, R.O., Sartori, D.J.M.: Drying of grass seeds in cross flow band dryer. In Proceedings of the International Drying Symposium, Halkidiki, Greece, vol. B, pp. 1350–1357 (1998)
25. Yang, W., Siebenmorgen, T.J., Jia, C.C., Howwell, T.A., Cnossen, A.G.: Cross-Flow drying of rough rice as mapped on its glass transition state diagram. In: Proceedings of the International Drying Symposium (IDS), Noordwijkerhout, The Netherlands, CD ROM (2000)
26. Siebenmorgen, W.Y., Jia, C.C., Howell, T.A., Cuossen, A.G.: Cross-flow drying of rough rice as mapped and its glass transition state diagram. In: Proceedings of the International Drying Symposium (IDS), Noordwijkerhout, The Netherlands, CD ROM (2000)
27. Farias, R.P.: Biological products drying simulation in crossflow dryer. Master Thesis (Mechanical Engineering), Federal University of Campina Grande, Campina Grande, Brazil (2003). (In portuguese)
28. Farias, R.P., Holanda, P.R.H., Lima, A.G.B.: Drying of grains in conveyor dryer and cross flow: a numerical solution using finite-volume method. Braz. J. Agro-ind. Prod. **6**(1), 1–16 (2004)
29. Fick, A.: Ueber diffusion. Annln Phys. **170**(1), 59–86 (1855)
30. Sun-Tak, Hwang, Kammermeyer, K.: Membranes in Separations. Wiley, New York (1975)
31. Shukla, K.N.: Diffusion Processes During Drying of Solids. World Scientific, Singapore (1990)

32. Lewis, W.K.: The rate of drying of solid materials. J. Ind. Eng. Chem. **13**, 427–432 (1921)
33. Sherwood, T.K.: The drying of solid, Pt. I. Ind. Eng. Chem. **21**, 12–16 (1929)
34. Sherwood, T.K.: The drying of solids, Pt. II. Ind. Eng. Chem. **21**, 976–980 (1929)
35. Sherwood, T.K.: The drying of solids, Pt. III, Mechanism of the drying of pulp and paper. Ind. Eng. Chem. **22**, 132–136 (1930)
36. Newman, A.B.: Diffusion and surface emission equation. Trans. AIChe. **27**, 203–220 (1931)
37. Fortes, M., Okos, M.R.: Drying theories: their bases and limitations as applied to foods and grains. In: Mujumdar, A.S. (ed.) Advances in Drying, vol. 1. Hemisphere Publishing Corporation, Washington (1980)
38. Parry, J.L.: Mathematical modelling and computer simulation of heat and mass transfer in agricultural grain drying. A review. J. Agri. Eng. Res. **32**, 1–29 (1985)
39. Jumah, R.Y., Mujumdar, A.S., Raghavan, G.S.V.: A mathematical model for constant and intermittent batch drying of grains in a novel rotating jet spouted bed. Drying Technol. **14**(3–4), 765–802 (1996)
40. Rossi, S.J.: Psicrometry. João Pessoa, FUNAPE (1987). p. 60 (In Portuguese)
41. Patankar, S.V.: Numerical Heat Transfer and Fluid Flow. Hemisphere Publishing Corporation, New York (1980)
42. Maliska, C.R.: Computational Heat Transfer and Fluid Mechanics. Ed. LTC, Rio de Janeiro, Brazil (1995). (In Portuguese)
43. Versteeg, H.K., Malalasekera, W.: An Introduction to Computational Fluid Dynamics: The Finite Volume Method. Prentice Hall, London (1995)
44. Santiago, D.C., Farias, R.P., Lima, A.G.B.: Modeling and simulation of cross flow band dryer: a finite-volume approach. In: Proceedings of the International Drying Symposium (IDS), Beijing, China, CD-ROM (2002)
45. Fortes, M.: A non-equilibrium thermodynamics approach to transport phenomena in capillary-porous media with special reference of grain and foods. PhD Thesis, Purdue University (1978)

Food Dehydration: Fundamentals, Modelling and Applications

João M. P. Q. Delgado and Marta Vázquez da Silva

Abstract Food dehydration is a preservation technique used by the man almost since ever. From the oldest times to nowadays, the dehydration of food is done in order to obtain a product with longer shelf life and minimum losses of physical, chemical and organoleptic characteristics when compared to the fresh material. The objective of this work is to present a brief explanation of the most used techniques, such as convective dehydration, microwave, vacuum and freeze dehydration, and a combinations of these techniques. A critical analyses of the variation of the moisture content of the food with the time of dehydration is presented. Several mathematical models that are classified as theoretical, semi-theoretical and empirical are discussed. An overview of the osmotic dehydration is also given, not as a dehydration technique by itself, but as pre-treatment that will improved the results obtained when the product is subjected to one of the mentioned techniques.

Keywords Dehydration · Food · Mathematical models

J. M. P. Q. Delgado (✉)
Laboratory of Building Physics (LFC), Civil Engineering Department,
Faculty of Engineering, University of Porto, Porto, Portugal
e-mail: jdelgado@fe.up.pt

M. V. da Silva
Instituto Universitário da Maia (ISMAI), Avenida Carlos Oliveira Campos,
4475-690 Avioso S. Pedro, Castêlo da Maia, Portugal
e-mail: mvazquez@fe.up.pt

M. V. da Silva
Faculdade de Engenharia, Centro de Estudos de Fenómenos de Transporte (CEFT),
Universidade do Porto, Rua Dr. Roberto Frias s/n, 4200-465 Porto, Portugal

J. M. P. Q. Delgado and A. G. Barbosa de Lima (eds.), *Transport Phenomena and Drying of Solids and Particulate Materials*, Advanced Structured Materials 48,
DOI: 10.1007/978-3-319-04054-7_4,

1 Introduction

The dehydration of food has been practiced ever since by man. The origin of food dehydration came from the pre-history, when man sun dried foods to sustain himself in off-season periods. Like almost all the technological advances, food dehydration suffered developments in particularly difficult times for humanity. Coming into the last century, in times of war, for example, has become particularly important to be able to develop techniques that increased the shelf life of food products. Drying processes not only tend to prevent the growth of microorganisms and bacteria in food, but also lead to the decrease of its weight and volume, making easier the transportation to the war zones. In addition, people were able to store supplies in larger quantities and for longer periods, addressing the difficulties caused by food rationing, hunger and diseases, typical of wartimes.

An important contribution to the develop of drying techniques was given by Sherwood [114] through his work "Drying of solids" where, for the first time, was given a scientific approach to a technique that until then was essentially a popular procedure. Drying then became considered as a unit operation different from the others. From this point, efforts have been made in order to develop new techniques and equipment for obtaining higher quality products, keeping or even improving the organoleptic properties of the food. Instant beverage powers, spices, dry soup mixes, coffee, milk, fruits are some examples of dehydrated foods that can be found easily nowadays.

Dehydration is a unit operation in which part or all of the water present in the material is removed by sublimation or evaporation resulting from the application of heat [16]. Drying is the process that thermally removes volatile substances, resulting in a solid with a low volatile content or even zero [82]. This second definition is no more than a generalization of the term dehydration. In these definitions are not included mechanical processes of water removal, such as filtration or centrifugation, for example.

The main objective of drying is to increase the lifetime of food materials. Reducing the moisture content, a reduction of water activity is achieved which reduces the enzymatic activity, inhibits the microbial growth and the progression of undesirable chemical reactions.

In some cases, drying means an improvement of the organoleptic properties of the food making it more desirable to the eyes of the consumer. Examples include dried fruit, vegetables, nuts, and many other snacks, as well as specific types of meat or fish.

Drying processes can also be used as a pre-treatment of foods, since separation processes, milling or mixing are often much simpler when the moisture content of the material is reduced.

As a natural consequence from a loss of water, the dried food have less weight and a smaller volume when compared to fresh one, which facilitates its packaging and transportation, factors that, from an economic point of view, are very important.

Dehydration processes can be classified as:

(i) Air convection dehydration. The heat is transferred, under atmospheric pressure, through the foodstuff either from heated air or from heated surfaces. The water vapour is removed with the air.
(ii) Microwave dehydration: The microwave radiation is thermally very efficient which reduces the energy requirements and the time of dehydration. Is commonly used coupled with other techniques.
(iii) Vacuum dehydration. In vacuum dehydration, the main advantage is taken of the fact that evaporation of water occurs more readily at lower pressures than at higher ones. Heat transfer in vacuum dehydration is generally by conduction, sometimes by radiation.
(iv) Freeze dehydration. In freeze dehydration, the water vapour is sublimed off frozen food. The food structure is better maintained under these conditions. Suitable temperatures and pressures must be established in the dryer to ensure that sublimation occurs.

It is also very common to particularize another dehydration process: the osmotic dehydration. According to principle of this technique, heat is not used to promote water removal. However, considering that is a unit operation that has as main objective to reduce moisture content, it is usual to classify the osmosis as a dehydration process. It is very often to submit the food material to osmotic dehydration treatments before being subjected to convective, vacuum or freezing techniques.

2 Dehydration Techniques

2.1 Air Convection Dehydration

Solar drying, also called open air drying, is, probably, the oldest and simplest convective drying technique, consisting in spreading the raw material over platforms exposed to sunlight. There is no doubt that is a simple and economical technique, however, presents some important constraints:

(i) there is no control over the drying rate and the drying process may result in loss of colour, germination, important nutritional changes or total spoilage of the food;
(ii) does not provide an uniform drying of the food, since it is very dependent on the way the raw material is spread on the drying platform;
(iii) if drying is too slow the growth of bacteria and fungi will prevent the food consumption;
(iv) environmental factors such as rain, dust, dirt, insects, birds and other animals are also decisive factors in food contamination and damage.

From an industrial point of view, this technique is not so interesting; since it is clear the need to control the drying process, however, all over the world, it is still practiced by popular.

In the industrial convective drying systems, an air stream is heated by an external power source and passes through the material, providing the required energy to promote the dehydration and transportation of the removed water. Dehydration involves more or less complex mechanisms of heat and mass transfer, that are highly conditioned by operating parameters, such as temperature (the greater the difference between the product and drying medium, the greater the rate of drying), humidity (higher humidity values means lower drying rates) or velocity of the air stream and by the product to be treated itself, such as surface (higher surfaces areas means higher rates of drying), size, shape, density, porosity, chemical composition (for example, the water that is most difficult to remove is that chemically bounded in the form of hydrates), etc.

In general, the process of convective drying can be divided into three main stages:

(i) first, the wet material (food) must be heated until the drying temperature, this is a very fast procedure and, most of the times, is not detected in the drying curves, being neglected; in this period the drying rate is essentially zero, and therefore the moisture content remains almost constant and equal to the initial moisture content;
(ii) secondly, it occurs the transport of the water present in the macro and micro pores of the material to the surface and subsequent evaporation; at this stage the drying rate is constant, which means the moisture content decreases linearly with the drying time;
(iii) finally, the drying rate starts to decrease until it reaches the final value; for the vast majority of the food, this is the most important period of the drying process, and the one that will define the final moisture content of the dried material.

In Fig. 1 it can be seen an example of a drying curve where the drying rate is plotted against time. In Fig. 2 the moisture content as a function of time is shown.

In Fig. 3 the temperature of the material during the drying process is plotted. Initially, the material is heated until the operating value. During the period where the drying rate is constant, the temperature of the material is essentially constant due to the effects of evaporative cooling, i.e., when the water is vaporized by a hot gas stream, the water film at the surface of the solid assume the wet bulb temperature of the drying gas, and the process can be considered as adiabatic, which means there is no heat exchange with the exterior. With the increasing of the drying time, during the falling drying rate period, because the moisture content of the material decreases, the effect of evaporative cooling is not as effective, resulting in an increase of the temperature of the material. This may be particularly important for heat sensitive foods [130].

Although the drying process is highly dependent on the characteristics of the material to be treated, it is possible to make a first classification of the materials as

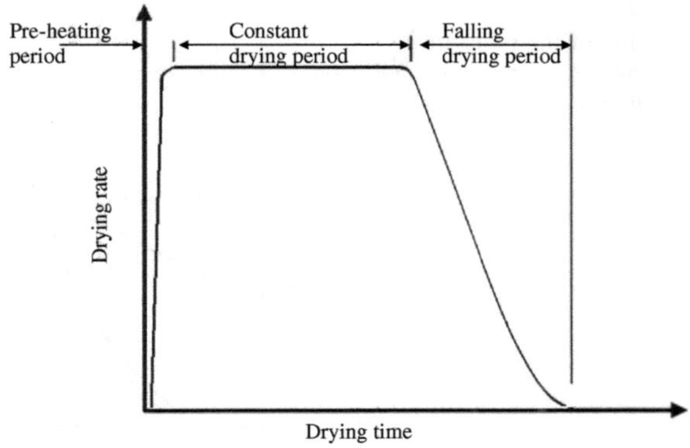

Fig. 1 Drying rate as a function of time for a typical drying curve

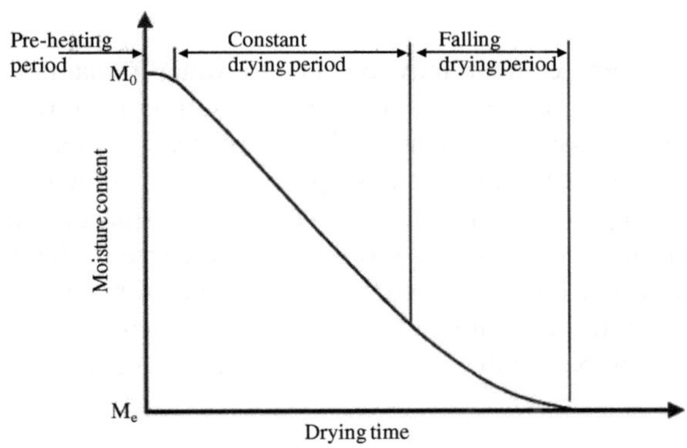

Fig. 2 Moisture content as a function of time for a typical drying curve

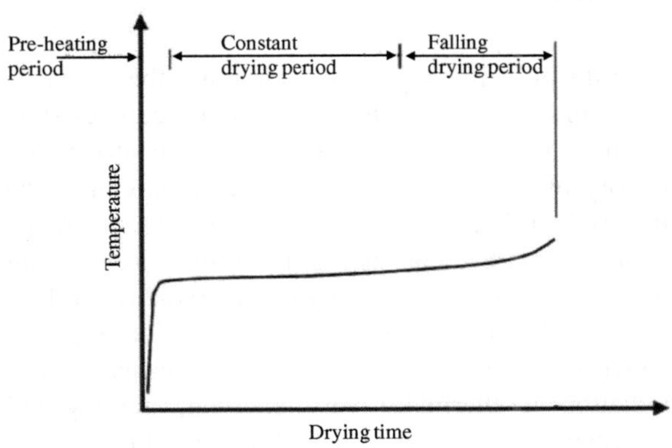

Fig. 3 Temperature of the material as a function of time for a typical drying curve

non-hygroscopic and hygroscopic. In the first case, after drying, the moisture content of the material is zero; for hygroscopic materials, representative of most of the foods, the drying threshold corresponds to a non-zero humidity value. The final

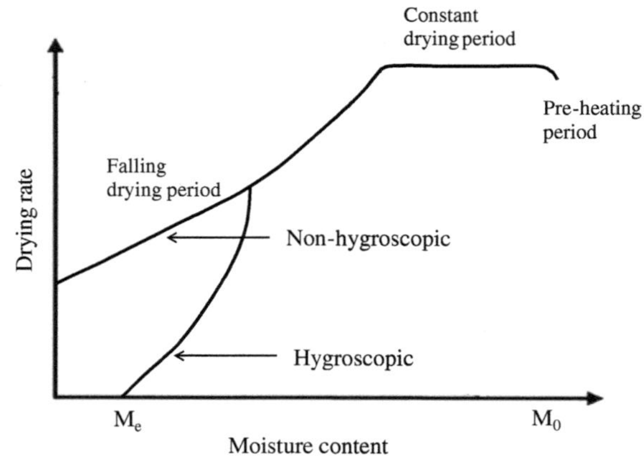

Fig. 4 Drying rate as a function of the moisture content for hygroscopic and non-hygroscopic materials

moisture content may be due to water remaining in the material due to the existence of closed capillaries, or resulting from the surface tension of water. The final moisture content of the hygroscopic material is referred to as equilibrium moisture content, M_e, corresponding to the point at which the vapor pressure of the water on the material is equal to the partial pressure of water in the stream of drying. In Fig. 4, it is shown a typical curve of the drying rate as a function of the moisture content for each of these kind of materials.

From Fig. 4 it can be seen that, for hygroscopic materials, the drying rate falls to zero before the sample is completely dried and the process finishes at this point.

2.2 Microwave Dehydration

In spite of being a simple technique, the convective food dehydration has the problem of being a highly energetic process, especially in the period in which the drying rate is decreasing. This is mainly the result of an increased of the complexity of the mechanisms of water removing, enhanced by the reduction of the material surface and its consequent shrinkage. As a consequence, the drying time is high and it becomes necessary to increase the temperature at which dehydration occurs, often resulting in thermal degradation of the material, leading to changes in color, taste, or nutrient modification. The application, in a combined form, of convective drying and microwave reduces these effects, since the microwave radiation is thermally very efficient, thereby reducing the drying time and increasing the quality of dehydrated product [15, 44, 73, 7, 64, 67, 77, 78, 113].

Typically the application of microwave is done using a constant heating power, and higher heating powers, mean higher heating rates and shorter drying times. However, because the non-uniform electric field distribution of the micro wave inside the material, the heating is not uniform, which, enhanced by a mass

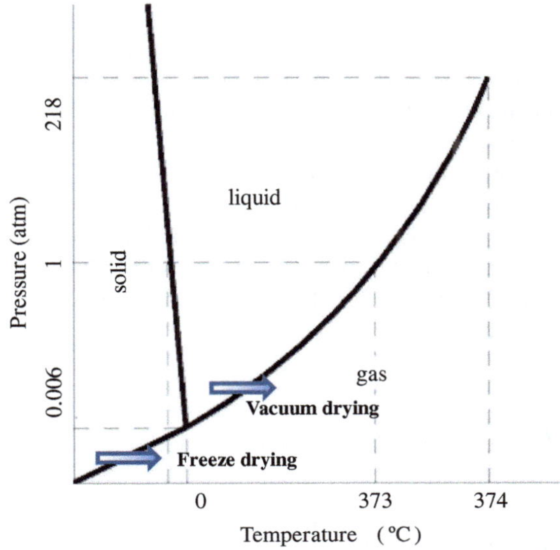

Fig. 5 Phase diagram of water

reduction of the material during the dehydration, may yield undesirable phenomena of overheating and charring of the food. To minimize this problem, it may be used two alternative ways of applying microwave radiation: pulsed microwave, where moments in which the dehydration process is accelerated by the use of microwave alternate with moments where it isn't done the application of radiation [47]; or providing controlled microwave power, preventing the product to reach high temperatures [3].

The drying kinetics of foods during microwave heat treatment has recently been a subject of interest for various investigators, being reported drying kinetics for macaroni [8], garlic [27, 112, 134], apple [13, 66, 141, 142], wheat [136], carrot [28, 121, 122, 138, 140], parsley [119], black tea [91], mushroom [102], lactose [79, 80, 139], cabbage [144], mint [87], okra [29, 30], spinach [31], basil [34], kiwi [77, 78], banana [76] and celery leaves [35].

2.3 Vacuum Dehydration

From the analysis of the water phase diagram (Fig. 5), it is evident that, for lower values of pressure, the temperature that must be applied to promote the evaporation of water is reduced. Thus, the use of operating pressures lower than the atmospheric, enables the dehydration process to take place at lower temperatures. This technique is commonly referred to as vacuum dehydration.

The vacuum dehydration is particularly important when it comes to remove water from thermo sensitive food, that usually are subject to partial or complete deterioration when exposed to high temperatures. Additionally, and because the heat requirements are lower, this is an energetically less demanding technique, beyond that the use of atmospheres with smaller quantities of air decreases the risk of occurring oxidation reactions during dehydration, and characteristics as color, texture or flavor of the food are improved. On the other way, this is a technique with higher installation and operational costs, mainly related to technical issues associated with the need to lower the pressure.

Several studies on vacuum and combined-vacuum techniques can be found in literature for fruit juices [39, 40, 65], strawberries [68], apple [108], tea [56, 118], carrot [26, 28], garlic [27], rice [135], cranberries [123, 148, 149], pumpkin [7, 83], etc.

2.4 Freezing Dehydration

Freeze drying, or lyophilization, is a drying process where the solvent, usually water, and/or suspension medium is crystallized at low temperature and then sprayed directly from the solid state through a sublimation process.

Freeze drying was introduced in the 1940s for the production of dry plasma and blood products. Then, antibiotics and biological materials were prepared on an industrial scale. With regard to food products, the development of low temperature drying processes was driven by the need to resolve issues related to the loss of flavors and aromas of food, present in the conventional drying process [32, 60].

Regarding other methods of dehydration, freeze drying has advantages such as greater product stability at room temperature, the easier rehydration of the product by addition of water, the preservation of morphological, biochemical and immunological properties. These aspects are related to the fact that this unit operation occurs at a low temperature and in an atmosphere poor in oxygen. Besides, the porous structure and the surface area of the material are remained. In addition, the final product is more crispy than when dried by convective air drying.

The main problem with freeze drying is associated with high costs of operation and maintenance, resulting in the most expensive technique of dehydration. Compared to other processes, the cost of freeze drying are up to twice the costs associated with conventional drying methods [43].

In a typical lyophilization, the process begins by frozen the product at a uniform temperature. This procedure must be very rapid in order to obtain a product with small crystals at surface and amorphous structure. Subsequently, the pressure of the lyophilizer is reduced, usually below 1 mbar, and the water is removed as vapor resulting from the sublimation of the ice crystals. In the foodstuff, the freezing temperature of water is lower than that of pure water and, due to the increasing concentration of the solutes present in the unfrozen product, there is a progressive decrease in freezing temperature throughout the process. The secondary drying stage is the longest part of the processing time, it is required to remove water that is tightly or chemically bound inside the product. Products are

Fig. 6 Freeze drying steps (Adapted from [81])

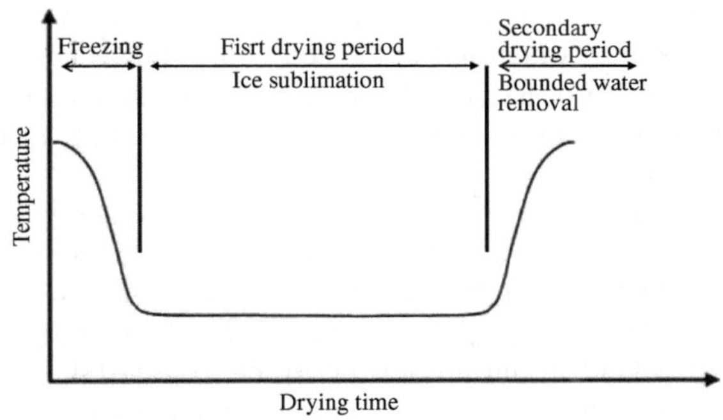

dried down to around 2 % moisture content. In Fig. 6 is plotted the temperature against time during a freeze dehydration process.

The most significant factors in freeze drying is the vacuum pressure, the flux of energy supplied to the material and the temperature of the condenser, being the optimum operating values highly dependent on the material to be treated. The initial drying rate is high due to low resistance to heat or mass transfer. However, the formation of a resistant layer around the frozen product, which acts as insulation material, will lead to a progressive decrease in the drying rate as the process progresses.

In addition to food products like coffea [105, 106], fruit juices [1, 9, 37], tropical fruits [74, 75], abalone [150], strawberries [49], apple [50] or carrot [14], freeze dehydration is suitable for other goods, including: flowers, microorganisms, pharmaceuticals, medical devices, cosmetics, etc.

3 Osmotic Dehydration

Osmotic dehydration, from a fundamental point of view, is not a drying technique, however it is commonly used as a pre-treatment, prior to another preservation technique like convective air drying, microwave drying or freezing.

Osmotic dehydration is a treatment that allows partial removal water in a foodstuff and/or the addition of solutes in a controlled manner, respecting the initial product quality. The process consists of introducing the foods in a hypertonic solution, controlling the operating conditions in order to favour in a greater or lesser degree the addition of solutes and the food dehydration.

In osmotic dehydration the most used solutes are sugar syrup for the treatment of fruits and sodium chloride or brine for vegetables treatment. This is a multi-component diffusion process, in which the water flows from fruits or vegetables to solution along and with some components of the food, such as vitamins, minerals, fruits acids, and others. The sugar and/or salt migrate towards the fruit or vegetable. The obtained products are classified as intermediate moisture food, with lower water activity when compared to the fresh products. This is due both to the removal of water and to the addition of humectants (sugars, sodium chloride). In spite the process does not reduce water activity sufficiently to completely hinder the proliferation of microorganisms, it extends, to some degree, the shelf life of the material.

Although there are some studies of osmotically dehydrated products performed with meat [45] or fish [101], because fruits and vegetables have a much higher moisture content, the majority of the published work are devoted to these kind of food, like, papaya [84], mandarin [125], tomato [6], apple [59, 90, 100], pineapple [95, 98, 115], mushrooms [129], chayote [103], banana [107], pepper [42, 124], tomato [36], apricots [41], carrot and pumpkin (Pan et al. [90], and many others.

The works of Ponting et al. [96], Jackson and Mohamed [55] and Islam and Flink [54] review by Yadav and Singh [145] pointed out some advantages of osmotic dehydration, namely: it is a low temperature water removal process and hence minimum loss of colour and flavour take place; enzymatic and oxidative browning is prevented as the fruit pieces are surrounded by sugar, thus making it possible to retain good colour; removal of acid and uptake of sugar by the fruit pieces results in a sweeter product than conventionally dried product; the textural quality of product is better after reconstitution; simple equipments are required for the process. The use of an osmotic pre-treatment before conventional drying processes reduces the energy demand, since water is partially removed without phase change. Estimates of energy consumption per kg of water removed for convective drying of foods were at least twice as much as needed for osmotic dehydration [70]. It also increases sugar to acid ratio, and improves the texture of the material and the stability of pigments during drying and storage [99]. Before freezing, an osmotic treatment may be applied with the objective of reducing refrigeration loads by water removal from the tissue and improving storage with the addition of cryoprotectants [23].

Despite osmotic dehydration being a good alternative to primarily remove water from food, one must pay attention to some disadvantages like the reduction in acidity level which reduces the characteristic taste of some products, or the sugar coating in some products which forces to a quick rinsing in water after the treatment. In terms of time, osmotic dehydration is a very slow process. Figure 7 shows a synopsis of the state of the art of the applications of osmotic dehydration as a pre-treatment in food processing.

4 Drying Kinetics

Simulation models of the drying process are used for designing new and/or improving existing drying systems, predicting the air flow over the product, or even for the control of the process (Karathanos and Belessiotis, 1997; Mathioulakis et al., 1998; Xia and Sun, 2002). A deep knowledge of the transfer parameters (heat and mass transfer, diffusion and thermal conductivity coefficients) and the

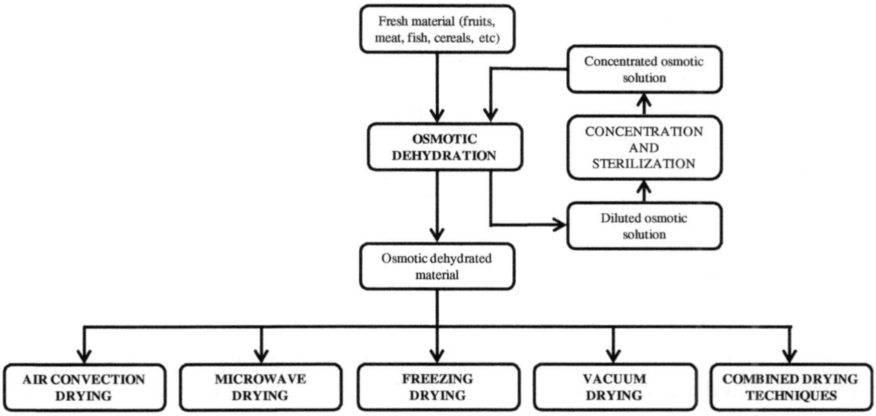

Fig. 7 Application of osmotic dehydration as a pre-treatment in food processing (Adapted from [120])

drying behaviour of the particular product to be dried are considered indispensable for the optimization of the drying performance used in simulation models. For this purpose, a sufficiently accurate model is required, capable of predicting the water removal rates and describing the drying performance of each particular product under the common conditions used in normal commercial relevant facilities [12].

The mathematical study of drying kinetics is a complex process since it involves not only mass, but also heat transfer phenomena that occurs simultaneously. As reported in Figs. 1 and 2, the drying process can be divided into three consecutive stages [5], where the first stage can be neglected, since it is very fast and characterized by drying rates close to zero, which means there is no impact in the moisture content of the material and, therefore, no meaning to the drying process.

4.1 Constant Drying Rate

In the constant drying rate period, the water is being evaporated from what is effectively a free water surface. The rate of removal of water can then be related to the rate of heat transfer, if there is no change in the temperature of the material and therefore all heat energy transferred to it must result in water evaporation. These two phenomena—heat and mass transfer—must predict the same rate of drying for a given set of circumstances. Therefore, the drying rate can be represented by Brooker et al. [17]

$$\frac{dM}{d\,\mathrm{t}} = \frac{h_c A}{h_{fg}}(T - T_{bu}) = \frac{h_m A}{R_v}\left(\frac{P_{vbu}}{T_{bu}} - \frac{P_{v\infty}}{T}\right) \tag{1}$$

where h_c is the global convective heat transfer coefficient, h_m is the global convective mass transfer coefficient, h_{fg} is the latent heat of vaporization, R_v is the universal water vapour constant (0.462 J kg^{-1} K^{-1}), A is the surface area, T is the drying temperature, T_{bu} is the wet bulb temperature, $P_{v\infty}$ is the vapour pressure, P_{vbu} is the vapour pressure for wet-bulb temperature and dM/dt is the (constant) drying rate.

4.2 Falling Drying Rate

The stage in which the drying rate is decreasing is the most important, since it corresponds to the period where the transport of water predominantly occurs. The thin-layer drying models, used to describe this period of the drying process, are divided into three main categories: theoretical, semi-theoretical and empirical models [109]. The major difference between these models is that the theoretical models suggest that the moisture transport is controlled mainly by internal resistance mechanisms, while the other two consider only external resistance. In the following sections it will be presented the most commonly used models of each category and several applications will be referred.

4.2.1 Pure Theorical Model: Fick's Second Law

It has been accepted that drying phenomenon of natural products during the falling rate stage is controlled by the mechanism of liquid and/or vapour diffusion. Assuming that the diffusion coefficient is constant, i.e., independent from the local moisture content, and that the material shrinkage is negligible, the Fick's second law, which establishes a relationship between the diffusion coefficient and the concentration gradient in a given environment, can be used

$$\frac{dM}{dt} = D_{eff} \frac{d^2M}{dr^2} \tag{2}$$

where M is the local moisture content, r is the diffusion path, t is the time and D_{eff} is the effective diffusion coefficient. In 1975, Crank [25] first gave the analytical solution for various geometries (sphere, cylinder, rectangular plate), later used by many researchers to describe the drying phenomena of various foodstuffs.

In the case of radial diffusion from a sphere of radius r_0, we have the following initial and boundary conditions:

$$\begin{cases} t = 0 \quad \text{and} \quad 0 \leq r < r_0, & M = M_0 \\ t > 0 \quad \text{and} \quad r = 0, & {}^{dM}/_{dr} = 0 \\ t > 0 \quad \text{and} \quad r = r_0, & M = M_e \end{cases} \tag{3}$$

where M_0 is the initial moisture content and M_e is the equilibrium moisture content. Those conditions states that moisture is initially uniformly distributed throughout the sample, the mass transfer is symmetric with respect to the central axis and the surface moisture content instantaneously reaches equilibrium with the surroundings. The analytical solution of Eq. (2) results in

$$MR = \frac{M - M_e}{M_0 - M_e} = \frac{6}{\pi^2}\sum_{m=1}^{\infty}\frac{1}{m^2}\exp\left[-\frac{m^2\pi^2 D_{eff} t}{r_0^2}\right] \tag{4}$$

where MR is the moisture ratio and $m = 1, 2, 3, \ldots$ is the number of terms taken into consideration.

Analogously, in the case of radial diffusion in a cylinder of radius r_0 and length L, we have the following initial and boundary conditions:

$$\begin{cases} t = 0 & \text{and} & 0 \le r < r_0, & M = M_0 \\ t > 0 & \text{and} & r = 0, & {dM}/{dr} = 0 \\ t > 0 & \text{and} & r = r_0, & M = \mathrm{M_e} \end{cases} \tag{5}$$

being the analytical solution of Eq. (2) given by

$$MR = \frac{M - M_e}{M_0 - M_e} = 4\sum_{m=1}^{\infty}\frac{1}{\lambda_n^2}\exp\left[-\frac{\lambda_m^2 D_{eff} t}{r_0^2}\right] \tag{6}$$

where λ_m are the Bessel equation roots.

Finally, for a infinite slab, the initial and boundary condition necessary to the analytical resolution of Eq. (2) are

$$\begin{cases} t = 0 & \text{and} & 0 \le r < L, & M = M_0 \\ t > 0 & \text{and} & r = 0, & {dM}/{dr} = 0 \\ t > 0 & \text{and} & r = L, & M = M_e \end{cases} \tag{7}$$

where and L is the half thickness of the slab for drying from both sides, or the thickness of the slab for drying from one side. The solution of Eq. (2) is then

$$MR = \frac{M - M_e}{M_0 - M_e} = \frac{8}{\pi^2}\sum_{n=1}^{\infty}\frac{1}{(2n-1)^2}\exp\left[-\frac{(2n-1)^2\pi^2 D_{eff} t}{4L^2}\right] \tag{8}$$

Assuming that only the first term in the series equations presented above is important [131] and applying the natural logarithm to both sides of equations, it is possible to obtain the effective diffusion coefficient from the slope of the straight line obtained when *ln(MR)* is plotted against time. For each of the considered geometries the corresponding equations are:

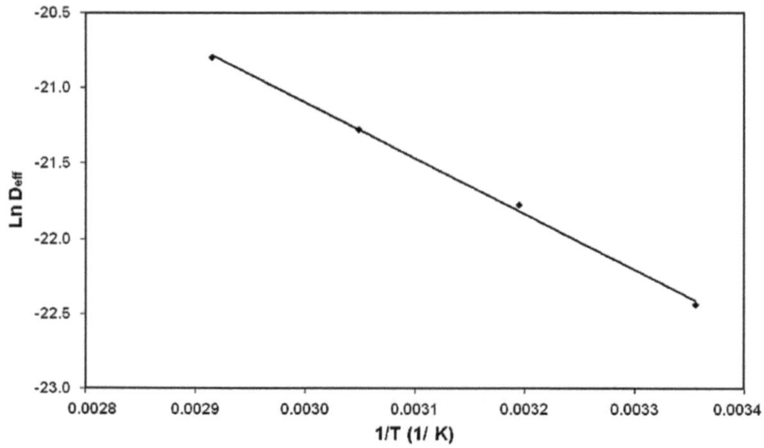

Fig. 8 Arrhenius-type relationship between effective diffusivity and temperature [61, 62]

$$\text{Sphere:} \quad \ln(MR) = \ln\left(\frac{6}{\pi^2}\right) - \frac{\pi^2 D_{eff}}{r_0^2} t \tag{9}$$

$$\text{Cylinder:} \quad \ln(MR) = \ln\left(\frac{4}{\lambda_1^2}\right) - \frac{\lambda_1^2 D_{eff}}{r_0^2} t \tag{10}$$

$$\text{Slab:} \quad \ln(MR) = \ln\left(\frac{8}{\pi^2}\right) - \frac{\pi^2 D_{eff}}{4L^2} t \tag{11}$$

The effect of temperature over the effective diffusion coefficient is an important factor to improve the quality of the predicted values of *MR*, and can be addressed through an Arrhenius relationship

$$D_{eff} = D_0 \exp\left(-\frac{E_a}{RT}\right) \tag{12}$$

where D_0 is a diffusivity constant equivalent to the diffusivity at infinitely high temperature, E_a is the activation energy and T is the absolute temperature. Plotting experimental values of $ln(D_{eff})$ as a function of $1/T$, it is obtained a straight line where the intercept gives D_0 and the slope gives $-E_a/R$. In Fig. 8 is presented an example of this procedure made by Kashaninejad et al. [61, 62].

4.2.2 Semi-Theoretical Models

The semi-theoretical models are derived directly from the general solution of Fick's law by simplification. It is important to note that the semi-theoretical models are valid only in the specific parameters range for which they were obtained, namely temperature, relative humidity, air flow velocity and moisture content.

The basic equation used to characterize the thin layer drying process, is similar to the Newton's law for cooling, and considers only one drying constant to describe the effect of all transport mechanisms existing during the drying process. Assuming that the bulk moisture content depends only on the time, the equation suggested by Lewis [71] has the form

$$MR = \exp(-kt) \tag{13}$$

To overcome the shortcomings of the exponential model, Page [89] proposed a model that includes an empirical modification in the time term by introducing an exponent n

$$MR = \exp(-kt^n) \tag{14}$$

He firstly applied this model to the shelled corn dehydration, and stated that the parameter k and n were air temperature and dew point dependent. Henderson and Pabis (1961) introduced a slight modification in the Lewis equation, by including a constant a in the exponential term

$$MR = a\exp(-kt) \tag{15}$$

In Overhults et al. [86] modelled the soybean dehydration using an approach to Page's model

$$MR = \exp\left[-(kt)^n\right] \tag{16}$$

A year later, Henderson (1974) proposed a two-term model

$$MR = a\exp(-k_0t) + b\exp(-k_1t) \tag{17}$$

An approach of the diffusion was suggested by Sharaf-Elden et al. [110, 111]

$$MR = a\exp(-k_0t) + (1-a)\exp(-k_1t) \tag{18}$$

In 1999, Karathanos used a modified Hander and Pabis equation to determine the water content of dried fruits

$$MR = a\exp(-k_0t) + b\exp(-k_1t) + c\exp(-k_2t) \tag{19}$$

Table 1 Semi-theoretical thin-layer drying models

Model	References
Lewis $MR = \exp(-kt)$	Carbonell et al. [19], Hansen et al. [51], Ayensu [10], Ozdemir and Devres [88], Panchariya et al. [91], Togrul and Pehlivan [127], Akgun and Doymaz [4], Akpinar [5], Babalis et al. [12], Kashaninejad et al. [61, 62], Upadhyay et al. [132]
Page $MR = \exp(-kt^n)$	White et al. [143], Chinnan [22], Karathanos and Belessiotis [57], Park et al. [92], Senadeera et al. (2003), Doymaz [38], Akgun and Doymaz [4], Akpinar [5], Babalis et al. [12], Kaleemullha and Kailappan (2006), Saeed et al. [104], Kashaninejad et al. [61, 62]
Henderson and Pabis $MR = a\exp(-kt)$	Henderson and Pabis [53], Sharaf-Elden et al. [110, 111], Chinnan [22], Ozdemir and Devres [88], Akgun and Doymaz [4], Akpinar [5], Babalis et al. [12], Saeed et al. [104], Kashaninejad et al. [61, 62]
Modified Page $MR = \exp[-(kt)^n]$	Sogi et al. [117], Ceylan et al. [21], Goyal et al. [46], Akgun and Doymaz [4], Akpinar [5], Babalis et al. [12]
Two-term $MR = a\exp(-k_0t) + b\exp(-k_1t)$	Sharaf-Elden et al. [110, 111], Rahman et al. [97], Yaldiz et al. [147], Dandamrongrak et al. [33], Togrul and Pehlivan [128], Lahsasni et al. [69], Akpinar [5], Babalis et al. [12], Kashaninejad et al. [61, 62], Wang et al. [141, 142]
Diffusion approach $MR = a\exp(-k_0t) + (1-a)\exp(-k_1t)$	Yaldiz and Ertekyn [146], Togrul and Pehlivan [128], Akgun and Doymaz [4], Akpinar [5], Wang et al. [141, 142]
Modified Henderson and Pabis $MR = a\exp(-k_0t) + b\exp(-k_1t) + c\exp(-k_2t)$	Karathanos [58], Yaldiz and Ertekyn [146], Akpinar [5], Kaya et al. [63]

In Table 1 are presented several studies that take into account the above Equations for modeling many different food materials. One must be aware that those equations are not the only ones used for drying modeling, but they are, probably, the most used ones.

4.3 Empirical Models

The empirical models directly correlate moisture content with time, having no physical connection with the drying process itself. Like the semi-theoretical models also purely empirical models are valid only for the experimental conditions which were developed.

Table 2 Empirical thin-layer drying models

Model	References
Thompson $t = a\ln(MR) + b[\ln(MR)]^2$	Thompson et al. [126], Paulsen and Thompson [93], Babalis et al. [12], Kashaninejad et al. [61, 62]
Wang and Singh $MR = 1 + at + bt^2$	Wang and Singh [137], Akgun and Doymaz [4], Akpinar [5], Babalis et al. [12]
Welbull distribution $MR = \exp\left[-\left(\frac{t}{k}\right)^n\right]$	Babalis et al. [12]

In 1968, Thompson et al. proposed an equation for shelled corn drying.

$$t = a\ln(MR) + b[\ln(MR)]^2 \tag{20}$$

In 1978, Wang and Sing applied an equation to rough rice.

$$MR = 1 + at + bt^2 \tag{21}$$

The standardised Weibull model can be applied in many biological systems, and it was found valuable in the modelling of dehydration–rehydration phenomena mainly for differentiating between diffusion and external resistance processes (Marabi et al., 2003; [24])

$$MR = \exp\left[-\left(\frac{t}{k}\right)^n\right] \tag{22}$$

In Table 2 are presented some examples where those empirical models were successful applied.

4.4 Determination of Desorption Isotherms

A sorption analysis must be performed to know the moisture content of the product at equilibrium, *Me*, since drying behavior is influenced by adsorption–desorption characteristics. This state of equilibrium results from multiple interactions on a microscopic scale, which is described by a relationship between the equilibrium water content of the product to be dried and the relative humidity of the atmosphere which surrounds it at a constant air temperature [11]. In literature it possible to find numerous isotherms equations that can be used to obtain the equilibrium moisture content of foodstuff, from which we selected those are listed hereafter:

- BET [18]

$$M = \frac{M_m\, C\, a_w}{(1 - a_w)\,(1 - a_w + C\, a_w)} \tag{23}$$

- Oswin [85]

$$M = A\left(\frac{a_w}{1 - a_w}\right)^B \quad (24)$$

- Smith [116]

$$M = A - B\ln(1 - a_w) \quad (25)$$

- Halsey [48]

$$M = \left[-\frac{A}{T\ln(a_w)}\right]^{1/B} \quad (26)$$

- Henderson [52]

$$M = \left[-\frac{\ln(1 - a_w)}{A}\right]^{1/B} \quad (27)$$

- Caurie [20]

$$M = \exp(A + Ba_w) \quad (28)$$

- Modified Henderson [126]

$$M = \left[-\frac{\ln(1 - a_w)}{A(T + C)}\right]^{1/B} \quad (29)$$

- GAB [133]

$$M = \frac{M_m\, C\, K\, a_w}{(1 - K\, a_w)\,(1 - Ka_w + C\, K\, a_w)} \quad (30)$$

- Modified BET (Aguirre et al., 1989)

$$M = \frac{M_m C a_w}{(1 - a_w)[1 - C\ln(1 - a_w)]} \quad (31)$$

- Peleg [94]

$$M = A\, a_w^C + B\, a_w^D \quad (32)$$

- Motta Lima (Lima et al. [72]

$$M = \frac{\ln\left(1 - a_w^C\right)}{A\exp\left(-\frac{B}{T}\right)} \tag{33}$$

5 Conclusions

The drying of food, in particular its dehydration, is a process that dates back to the earliest times. Its application at industrial level is becoming more widespread, and the study and dissemination of information on this unit operation is useful. This chapter is an approach of the main techniques used in food dehydration, stating in a summary form, the operating principle and the main advantages and constraints. Throughout the text references for further reading are mentioned.

The study of physical and structural changes during dehydration is important because they are related with different aspects of food processing, as well as with the quality of the dehydrated product and its acceptance by the consumer. It is clear that the use of procedures that minimize the changes of structure, colour, organoleptic, physical and chemical properties are critical to the final quality of the dried product, although often this means higher costs of operation and/or maintenance, as in the case of freeze dehydration.

The osmotic dehydration, as a pre-treatment, is a very commonly used technique, because it reduces the moistures content of the fresh product and the conventional dehydration process is easier, less aggressive to the material and requires smaller amounts of energy, being less expensive.

The mathematical study of drying kinetics is a complex process since it involves not only mass, but also heat transfer phenomena that occur simultaneously, nevertheless the knowledge about simulation models of the drying process are of decisive importance for designing new and/or improving existing drying systems. Theoretical, semi-theoretical and empirical models to describe the drying rate and/or moisture content as a function of the operating parameters are presented here, together with the most used isotherms equations necessary to predict the equilibrium moisture content.

Acknowledgments J.M.P.Q. Delgado would like to thank Fundação para a Ciência e a Tecnologia (FCT) for financial support through the grant SFRH /BPD /84377 /2012.

Nomenclature

A	Surface area
a	Thin-layer drying model parameter
A	Isotherm parameter
a_w	Water activity

b	Thin-layer drying model parameter
B	Isotherm parameter
C	Isotherm parameter
D	Isotherm parameter
D_0	Diffusivity constant equivalent to the diffusivity at infinitely high temperature
D_{eff}	Effective diffusion coefficient
dM/dt	Drying rate
E_a	Activation energy
h_c	Global convective heat transfer coefficient
h_{fg}	Latent heat of vaporization
h_m	Global convective mass transfer coefficient
k	Thin-layer drying model parameter
K	Isotherm parameter
k_0	Thin-layer drying model parameter
k_1	Thin-layer drying model parameter
k_2	Thin-layer drying model parameter
L	Length (of the cylinder, the half thickness of the slab for drying from both sides, or the thickness of the slab for drying from one side)
M	Local moisture content
M_0	Initial moisture content
M_e	Equilibrium moisture content
M_m	Monolayer moisture content
MR	Moisture ratio
n	Thin-layer drying model parameter
P_{vbu}	Vapour pressure for wet
$P_{v\infty}$	Vapour pressure
r	Diffusion path
r_0	Radius (of the sphere or the cylinder)
R_v	Universal water vapour constant (0.462 $J.kg^{-1}.K^{-1}$)
T	Drying temperature
t	Drying time
T_{bu}	Wet bulb temperature
λ_m	Bessel equation roots

References

1. Acevedo, B.A., Chaves, M.G., Avanza, M.V., Dellacassa, E.S.: Freeze-drying concentration of Rangpur lime juice. Int. J. Food Sci. Technol. **49**(2), 423–428 (2014)
2. Aguerre, R.J., Suarez, C., Viollaz, P.E.: New BET type multiplayer sorption isotherms. Part II: modelling water sorption in foods. Lebensmittel-Wissenchaft und Technol. **22**, 192–195 (1989)

3. Ahrnéa, L.M., Pereirab, N.R., Staacka, N., Floberga, P.: Drying Technol. **25**(7–8), 1149–1153 (2007)
4. Akgun, N.A., Doymaz, I.: Modelling of olive cake thin-layer drying process. J. Food Eng. **68**, 445–461 (2005)
5. Akpinar, E.K.: Determination of suitable thin layer drying curve model for some vegetables and fruits. J. Food Eng. **73**(1), 75–84 (2006)
6. Al-Harahsheh, M., Al-Muhtaseb, A.H., Magee, T.R.A.: Microwave drying kinetics of tomato pomace: effect of osmotic dehydration. Chem. Eng. Process. **48**(1), 524–531 (2009)
7. Alibas, I.: Microwave, air and combined microwave-air-drying parameters of pumpkin slices. LWT Food Sci. Technol. **40**(8), 1445–1451 (2007)
8. Altan, A., Maskan, M.: Microwave assisted drying of short-cut (ditalini) macaroni:drying characteristics and effect of drying processes on starch properties. Food Res. Int. **38**, 787–796 (2005)
9. Auleda, J.M., Raventós, M., Sánchez, J., Hernández, E.: Estimation of the freezing point of concentrated fruit juices for application in freeze concentration. J. Food Eng. **105**(2), 289–294 (2011)
10. Ayensu, A.: Dehydration of food crops using a solar dryer with convective heat flow. Sol. Energy **59**, 121–126 (1997)
11. Azzouz, S., Guizani, A., Joma, W., Belghith, A.: Moisture diffusivity and drying kinetic equation of convective drying of grapes. J. Food Eng. **55**(4), 323–330 (2002)
12. Babalis, S.J., Papanicolaou, E., Kyriakis, N., Belessiotis, V.G.: Evaluation of thin-layer drying models for describing drying kinetics of figs (Ficus carica). J. Food Eng. **75**, 205–214 (2006)
13. Bilbao-Sáinz, A., Andrés, C., Chiralt, A., Fito, P.: Microwaves phenomena duringdrying of apple cylinders. J. Food Eng. **74**(1), 160–167 (2006)
14. Boeh-Ocansey, O.: Some factors influencing the freeze drying of carrot discs in vacuo and at atmospheric pressure. J. Food Eng. **4**(3), 229–243 (1985)
15. Bouraoui, M., Richard, P., Durance, T.: Microwave and convective drying of potato slices. J. Food Process Eng. **17**, 353–363 (1994)
16. Brennan, J.G.: Food Dehydration: A Dictionary and Guide. Butterworth-Heinemann Ltd, Oxford (1994). ISBN 978-0750611305
17. Brooker, D.B., Bakker-Arkema, F.W., Hall, C.W.: Drying and Storage of Grains and Oilseeds. The AVI Publishing Company, Westport (1992)
18. Brunauer, S., Emmett, P.H., Teller, E.: Adsorption of gases in multimolecular layers. J. Am. Chem. Soc. **60**, 309 (1938)
19. Carbonell, J.V., Pinaga, F., Yusa, V., Pena, J.L.: Dehydration of paprika and kinetics of color degradation. J. Food Eng. **5**(3), 179–193 (1986)
20. Caurie, M.: A new model equation for predicting safe storage moisture levels for optimum stability of dehydrated foods. J. Food Technol. **5**, 301–307 (1970)
21. Ceylan, I., Aktas, M., Doğan, H.: Mathematical modeling of drying characteristics of tropical fruits. Appl. Therm. Eng. **27**, 1931–1936 (2007)
22. Chinnan, M.S.: Evaluation of selected mathematical models for describing thin layer drying of in-shell pecans. Trans. ASAE **27**(2), 610–615 (1984)
23. Chiralt, A., Martínez-Navarrete, N., Martínez-Monzó, J., Talens, P., Moraga, G., Ayala, A., Fito, P.: Changes in mechanical properties throughout osmotic processes: cryoprotectant effect. J. Food Eng. **49**(2–3), 129–135 (2001)
24. Corzo, O., Bracho, N., Pereira, A., Vásquez, A.: Weibull distribution for modeling air drying of coroba slices. LWT Food Sci. Technol. **41**(10), 2023–2028 (2008)
25. Crank, J.: The Mathematics of Diffusion, 2nd edn. Oxford University Press, London (1975)
26. Cui, Z.-W., Xu, S.-Y., Sun, D.-W.: Effect of microwave-vacuum drying on the carotenoids retention of carrot slices and chlorophyll retention of Chinese chive leaves. Dry. Technol. **22**(3), 563–575 (2004)
27. Cui, Z.-W., Xu, S.-Y., Sun, D.-W.: Dehydration of garlic slices by combined microwave-vacuum and air drying. Dry. Technol. **21**(7), 1173–1184 (2003)

28. Cui, Z.-W., Xu, S.-Y., Sun, D.-W., Chen, W.: Temperature changes during microwave-vacuum drying of sliced carrots. Dry. Technol **23**(5), 1057–1074 (2005)
29. Dadali, G., Apar, D.K., Ozbek, B.: Microwave drying kinetics of okra. Dry. Technol **25**(5), 917–924 (2007)
30. Dadali, G., Apar, D.K., Ozbek, B.: Estimation of effective moisture diffusivity of okra for microwave drying. Dry. Technol **25**(9), 1445–1450 (2007)
31. Dadali, G., Demirhan, E., Ozbek, B.: Microwave heat treatment of spinach: drying kinetics and effective moisture diffusivity. Dry. Technol **25**(10), 1703–1712 (2007)
32. Dalgleish, J.McN: Freeze-Drying. In the Food Industry. Elsevier, London (1990)
33. Dandamrongrak, R., Young, G., Mason, R.: Evaluation of various pre-treatment for dehydration of banana and selection of suitable drying models. J. Food Eng. **55**, 139–146 (2002)
34. Demirhan, E., Özbek, B.: Microwave drying characteristics of basil. J. Food Process. Preserv. **34**(3), 476–494 (2010)
35. Demirhan, E., Özbek, B.: Thin-layer drying characteristics and modeling of celery leaves undergoing microwave treatment. Chem. Eng. Comm. **198**, 957–975 (2011)
36. Dermesonlouoglou, E.K., Giannakourou, M.C., Taoukis, P.: Stability of dehydrofrozen tomatoes pretreated with alternative osmotic solutes. J. Food Eng. **78**(1), 272–280 (2007)
37. Deshpande, S.S., Cheryan, M., Sathe, S.K., Salunkhe, D.K.: Freeze concentration of fruit juices. Crit. Rev. Food Sci. Nutr. **20**(3), 173–248 (1984)
38. Doymaz, I.: Convective air drying characteristics of thin layer carrots. J. Food Eng. **61**, 359–364 (2004)
39. Drouzas, A.E., Tsami, E., Saravacos, G.D.: Microwave/vacuum drying of model fruit gels. J. Food Eng. **39**(2), 117–122 (1999)
40. Drouzas, A.E., Schubert, H.: Microwave application in vacuum drying of fruits. J. Food Eng. **28**(2), 203–209 (1996)
41. Fahloul, D., Lahbari, M., Benmoussa, H., Mezdour, S.: Effect of osmotic dehydration on the freeze drying kinetics of apricots. J. Food Agric. Environ. **7**(2), 117–121 (2009)
42. Falade, K.O., Oyedele, O.O.: Effect of osmotic pretreatment on air drying characteristics and colour of pepper (*Capsicum* spp) cultivars. J. Food Sci. Technol. **47**(5), 488–495 (2010)
43. Flink, J.: Energy analysis in dehydration processes. Food Technol. **31**, 77–78 (1977)
44. Garcia, R., Leal, F., Rolz, C.: Drying of bananas using microwave and air ovens. Int. J. Food Sci. Technol. **23**, 73–80 (1988)
45. Gerelt, B., Ikeuchi, Y., Suzuki, A.: Meat tenderization by proteolytic enzymes after osmotic dehydration. Meat Sci. **56**(3), 311–318 (2000)
46. Goyal, R.K., Kingsly, A.R.P., Manikantan, M.R., Ilyas, S.M.: Mathematical modeling of thin layer drying kinetics of plum in a tunnel dryer. J. Food Eng. **79**, 176–180 (2007)
47. Gunasekaran, S.: Grain drying using continuous and pulsed microwave energy. Dry. Technol. **8**(5), 1039–1047 (1990)
48. Halsey, G.: Physical adsorption on non-uniform surfaces. J. Chem. Phys. **16**, 931–937 (1948)
49. Hammami, C., René, F.: Determination of freeze-drying process variables for strawberries. J. Food Eng. **32**(2), 133–154 (1997)
50. Hammami, C., René, F., Marin, M.: Process-quality optimization of the vacuum freeze-drying of apple slices by the response surface method. Int. J. Food Sci. Technol. **23**(2), 145–160 (1999)
51. Hansen, R.C., Keener, H.M., ElSohly, H.N.: Thin-layer drying of cultivated taxus clippings. Trans. ASAE **36**(6), 1873–1877 (1993)
52. Henderson, S.M.: A basic concept of equilibrium moisture. Agric. Eng. **33**, 29–32 (1952)
53. Henderson, S.M., Pabis, S.: Grain drying theory I. Temperature effect on drying coefficient. J. Agric. Eng. Res. **6**(3), 169–174 (1969)
54. Islam, M.N., Flink, J.M.: Dehydration of potato II. Osmotic concentration and its effect on air drying behavior. J. Food Technol. **17**, 387–403 (1982)

55. Jackson, T.H., Mohamed, B.B.: The shambat process: new development arising from the osmotic dehydration of fruits and vegetables. Sudan J. Food Sci. Technol. **3**, 18–22 (1971)
56. Jeni, K., Yapa, M., Rattanadecho, P.: Design and analysis of the commercialized drier processing using a combined unsymmetrical double-feed microwave and vacuum system (case study: tea leaves). Chem. Eng. Process. **49**(4), 389–395 (2010)
57. Karathanos, V.T., Belessiotis, V.G.: Application of a thin layer equation to drying data of fresh and semi-dried fruits. J. Agric. Eng. Res. **74**, 355–361 (1999)
58. Karathanos, V.T.: Determination of water content of dried fruits by drying kinetics. J. Food Eng. **39**, 337–344 (1999)
59. Karathanos, V.T., Kostaropoulos, A.E., Saravacos, G.D.: Air-drying kinetics of osmotically dehydrated fruits. Dry. Technol. **13**(5–7), 1503–1521 (1995)
60. Karel, M.: Heat and mass transfer in freeze drying. In: Goldblith, S.A., Rey, L., Rothmayr, W.W. (eds.) Freeze Drying and Advanced Food Technology. Academic Press, London (1975). ISBN 0-12-288450-7
61. Kashaninejad, M., Mortazavi, A., Safekordi, A., Tabil, L.G.: Thin-layer drying characteristics and modeling of pistachio nuts. J. Food Eng. **78**, 98–108 (2007)
62. Kashaninejad, M., Mortazavi, A., Safekordi, A., Tabil, L.G.: Thin-layer drying characteristics and modeling of pistachio nuts. J. Food Eng. **78**, 98–108 (2007)
63. Kaya, A., Aydın, O., Demirtas, C.: Drying Kinetics of Red Delicious Apple. Biosyst. Eng. **96**(4), 517–524 (2007)
64. Khraisheha, M.A.M., McMinnb, W.A.M., Mageeb, T.R.A.: Quality and structural changes in starchy foods during microwave and convective drying. Food Res. Int. **37**(5), 497–503 (2004)
65. Kiranoudis, C.T., Tsami, E., Maroulis, Z.B.: Microwave vacuum drying kinetics of some fruits. Drying Technology **15**(10), 2421–2440 (1997)
66. Krokida, M.K., Maroulis, Z.B., Saravacos, G.D.: The effect of the method of drying on the color of dehydrated products. Int. J. Food Sci. Technol. **36**(1), 53–59 (2001)
67. Krokidaa, M.K., Maroulisa, Z.B.: Effect of microwave drying on some quality properties of dehydrated products. Dry. Technol. **17**(3), 449–466 (1999)
68. Krulis, M., Kühnert, S., Leiker, M., Rohm, H.: Influence of energy input and initial moisture on physical properties of microwave-vacuum dried strawberries. Eur. Food Res. Technol. **221**(6), 803–808 (2005)
69. Lahsasni, S., Kouhila, M., Mahrouz, M., Jaouhari, J.T.: Drying kinetics of prickly pear fruit (Opuntia ficus indica). J. Food Eng. **61**, 173–179 (2004)
70. Lewicki, P.P., Lenart, A.: Handbook of Industrial Drying, vol. 1, 2nd edn, p. 691. Marcel Dekker Inc, New York (1995)
71. Lewis, W.K.: The rate of drying of solid materials. Ind. Eng. Chem. **13**, 427–432 (1921)
72. Lima, O.C.M., Machado, G.D., Lucheis, R.M., Pereira, N.C.: Moisture equilibrium isotherms for a handmake kraft paper. In: Proceedings of the International Drying Symposium, São Paulo. CD-Rom (2004)
73. Lin, T.M., Durance, T.D., Scaman, C.H.: Characterization of vacuum microwave, air and freeze dried carrot slices. Food Res. Int. **31**(2), 111–117 (1998)
74. Marques, L.G., Ferreira, M.C., Freire, J.T.: Freeze-drying of acerola (Malpighia glabra L.). Chem. Eng. Process. **46**(5), 451–457 (2007)
75. Marques, L.G., Freire, J.T.: Analysis of freeze-drying of tropical fruits. Dry. Technol. **23**(9–11), 2169–2184 (2005)
76. Maskan, M.: Microwave/air and microwave finish drying of banana. J. Food Eng. **44**(2), 71–78 (2000)
77. Maskan, M.: Kinetics of colour change of kiwifruits during hot air and microwave drying. J. Food Eng. **48**(2), 169–175 (2001)
78. Maskan, M.: Drying, shrinkage and rehydration characteristics of kiwifruits during hot air and microwave drying. J. Food Eng. **48**(2), 177–182 (2001)
79. McMinn, W.A.M.: Prediction of moisture transfer parameters for microwave drying of lactose powder using Bi–G drying correlation. Food Res. Int. **37**, 1041–1047 (2004)

80. McMinn, W.A.M.: Thin-layer modeling of the convective, microwave, microwave-convective and microwave-vacuum drying of lactose powder. J. Food Eng. **72**(2), 113–123 (2006)
81. Mellor, J.D.: Fundamentals of Freeze-Drying. Academic Press, London (1978)
82. Mujumdar, A.S. (ed.): Handbook of Industrial Drying, 2nd edn. New York, Marcel Dekker (1995). ISBN 0-8247-9644-6
83. Nawirska, A., Figiel, A., Kucharska, A.Z., Sokół-Łetowska, A., Biesiada, A.: Drying kinetics and quality parameters of pumpkin slices dehydrated using different methods. J. Food Eng. **54**(1), 14–20 (2009)
84. Nimmanpipug, N., Therdthai, N., Dhamvithee, P.: Characterisation of osmotically dehydrated papaya with further hot air drying and microwave vacuum drying. Int. J. Food Sci. Technol. **48**(6), 1193–1200 (2013)
85. Oswin, C.R.: The kinetics of package life III. The Isotherm J. Chem. Ind. **65**, 419–421
86. Overhults, D.G., White, H.E., Hamilton, H.E., Ross, I.J.: Drying soybean with heated air. Trans. ASAE **16**, 112–113 (1973)
87. Özbek, B., Dadali, G.: Thin-layer drying characteristics and modeling of mint leaves undergoing microwave treatment. J. Food Eng. **83**(4), 541–549 (2007)
88. Ozdemir, M., Devres, Y.O.: The thin layer drying characteristics of hazelnuts during roasting. J. Food Eng. **42**, 225–233 (1999)
89. Page, G.E.: Factors influencing the maximum rates of air drying shelled corn in thin layers. M. Sc. thesis, Purdue University (1949)
90. Pan, Y.K.A., Zhao, L.J.A., Zhang, Y.A., Chen, G.B., Mujumdar, A.S.: Osmotic dehydration pretreatment in drying of fruits and vegetables. Dry. Technol. **21**(6), 1101–1114 (2003)
91. Panchariya, P.C., Popovic, P.C., Sharma, A.L.: Thin-layer modelling of black tea drying process. J. Food Eng. **52**(4), 349–357 (2002)
92. Park, K.J., Vohnikova, Z., Brod, F.P.R.: Evaluation of drying parameters and desorption isotherms of garden mint leaves (Mentha crispa L.). J. Food Eng. **51**, 193–199 (2002)
93. Paulsen, M.R., Thompson, T.L.: Drying endysus of grain sorghum. Trans. ASAE **16**, 537–540 (1973)
94. Peleg, M.: Assessment of a semi-empirical four parameter general model for sigmoid moisture sorption isotherms. J. Food Process Eng. **16**, 21–37 (1993)
95. Pokharkar, S.M., Prasad, S.: Air drying behaviour of osmotically dehydrated pineapple. J. Food Sci. Technol. **39**(4), 384–387 (2002)
96. Ponting, J.D., Watters, G.G., Forrey, R.R., Jackson, R., Stanley, W.L.: Osmotic dehydration of fruits. Food Technol. **20**, 125–128 (1966)
97. Rahman, M.S., Perera, C.O., Theband, C.: Desorption isotherm and heat pump drying kinetics of peas. Food Res. Int. **30**, 485–491 (1998)
98. Rahman, S., Lamb, J.: Air drying behavior of fresh and osmotically dehydrated pineapple. J. Food Process Eng. **14**(3), 163–171 (1991)
99. Raoult-Wack, A.L.: Recent advances in the osmotic dehydration of foods. Trends Food Sci. Technol. **5**(8), 255–260 (1994)
100. Reppa, A., Mandala, J., Kostaropoulos, A.E., Saravacos, G.D.: Influence of solute temperature and concentration on the combined osmotic and air drying. Dry. Technol. **17**(7–8), 1449–1458 (1999)
101. Reyes, M.G., Corzo, O., Bracho, N.: Optimization of the osmotic dehydration of sardines by response surface methodology. Revista Cientifica de la Facultad de Ciencias Veterinarias de la Universidad del Zulia **15**(4), 377–384 (2005)
102. Rodríguez, R., Lombraña, J.I., Kamel, M., Elvira, C.: Kinetic and quality study of mushroom drying under microwave and vacuum. Dry. Technol. **23**(9–11), 2197–2213 (2005)
103. Ruiz-López, I.I., Huerta-Mora, I.R., Vivar-Vera, M.A., Martínez-Sánchez, C.E., Herman-Lara, E.: Effect of osmotic dehydration on air-drying haracteristics of chayote. Dry. Technol. **28**(10), 1201–1212 (2010)

104. Saeed, I.E., Sopian, K., Zainol Abidin, Z.: Drying kinetics of Roselle (*Hibiscus sabdariffa* L.): dried in constant temperature and humidity chamber. In: Muchtar (ed.) Proceedings of SPS 2006, pp. 143–148. Permata, Bangi, S.D.E., Malaysia, 29–30 Aug 2006
105. Sagara, Y., Kaminishi, K., Goto, E., Watanabe, T., Imayoshi, Y., Iwabuchi, H.: Characteristic evaluation for volatile components of soluble coffee depending on freeze-drying conditions. Dry. Technol. **22**(9–11), 2185–2196 (2005)
106. Sagara, Y., Ichiba, J-i: Measurement of transport properties for the dried layer of coffee solution undergoing freeze drying. Dry. Technol. **12**(5), 1081–1103 (1994)
107. Sankat, C.K., Castaigne, F., Maharaj, R.: The air drying behaviour of fresh and osmotically dehydrated banana slices. Int. J. Food Sci. Technol. **31**(2), 123–135 (1996)
108. Sham, P.W.Y., Scaman, C.H., Durance, T.D.: Texture of vacuum microwave dehydrated apple chips as affected by calcium pretreatment, vacuum level, and apple variety. J. Food Sci. **66**(9), 1341–1347 (2001)
109. Sharaf-Eldeen, Y.I., Hamdy, M.Y.: Falling rate drying of fully exposed biological materials: a review of mathematical models. ASAE Paper No. 79-6622. 1979 Winter Meeting of ASAE (1979)
110. Sharaf-Elden, Y.I., Blaisdell, J.L., Hamdy, M.Y.: A model for ear corn drying. Trans. ASAE **23**(5), 1261–1265, 1271 (1980)
111. Sharaf-elden, Y.I., Blaisdell, J.L., Hamdy, M.Y.: A model for ear corn drying. Trans. ASAE **5**(4), 1261–1265 (1980)
112. Sharma, G.P., Prasad, S.: Optimization of process parameters for microwave drying of garlic cloves. J. Food Eng. **75**(4), 441–446 (2005)
113. Sharma, G.P., Prasad, S.: Comparison of convective and microwave—convective drying of garlic: kinetics and energy consumption. J. Food Sci. Technol. **39**(6), 603–608 (2002)
114. Sherwood, T.K.: The drying of solids. Ind. Eng. Chem. **21**(1), 12–16 (1929)
115. Shi, X.Q., Fito, P.: Vacuum osmotic dehydration of fruits. Dry. Technol. **11**(6), 1429–1442 (1993)
116. Smith, S.E.: The sorption of water vapour by high polymers. J. Am. Chem. Soc. **69**, 646 (1947)
117. Sogi, D.S., Shivhare, U.S., Garg, S.K., Bawa, S.A.: Water sorption isotherms and drying characteristics of tomato seeds. Biosyst. Eng. **84**(3), 297–301 (2003)
118. Someswararao, C., Srivastav, P.P.: A novel technology for production of instant tea powder from the existing black tea manufacturing process. Innovative Food Sci. Emerg. Technol. **16**, 143–147 (2012)
119. Soysal, Y.: Microwave drying characteristics of parsley. Biosyst. Eng. **89**(2), 167–173 (2004)
120. Spiess, W.E.L., Beshnilian, D.: Osmotic treatments in food processing. Current state and texture needs. In: Akritidis, C.B., Marinos-Kouris, D., Saravacos, G.D. (eds.) Proceedings of the 11th International Drying Symposium (IDS'98), vol A, pp. 47–56. Ziti Editions, Thessloniki (1998)
121. Stanisławski, J.: Drying of diced carrot in a combined microwave–fluidized bed dryer. Dry. Technol. **23**(8), 1711–1721 (2005)
122. Sumnu, G., Turabi, E., Oztop, M.: Drying of carrots in microwave and halogen lamp-microwave combination ovens. Lebensm.-Wiss. Technol. **38**(5), 549–553 (2005)
123. Sunjka, P.S., Rennie, T.J., Beaudry, C., Raghavan, G.S.V.: Microwave-convective and microwave-vacuum drying of cranberries: A comparative study. Dry. Technol. **22**(5), 1217–1231 (2004)
124. Swain, S., Samuel, D.V.K., Bal, L.M., Kar, A., Sahoo, G.P.: Modeling of microwave assisted drying of osmotically pretreated red sweet pepper (*Capsicum annum* L.). Food Sci. Biotechnol. **21**(4), 969–978 (2012)
125. Therdthai, N., Zhou, W., Pattanapa, K.: Microwave vacuum drying of osmotically dehydrated mandarin cv. (Sai-Namphaung). Int. J. Food Sci. Technol. **46**(11), 2401–2407 (2011)
126. Thompson, T.L., Peart, R.M., Foster, G.H.: Mathematical Simulation of Corn Drying - A New Model. Transactions of the ASABE. **11**(4), 582–586 (1968)

127. Togrul, I.T., Pehlivan, D.: Modeling of thin layer drying kinetics of some fruits under open air sun drying process. J. Food Eng. **65**(3), 413–425 (2004)
128. Togrul, I.T., Pehlivan, D.: Mathematical modeling of solar drying of apricots in thin layers. J. Food Eng. **55**(3), 209–216 (2002)
129. Torringa, E., Esveld, E., Scheewe, I., Van Den Berg, R., Bartels, P.: Osmotic dehydration as a pre-treatment before combined microwave-hot-air drying of mushrooms. J. Food Eng. **49**(2–3), 185–191 (2001)
130. Traub, D.A. http://www.process-heating.com/ext/resources/PH/Home/Files/PDFs/1002ph_dryingfiles.pdf (2014)
131. Tutuncu, M.A., Labuza, T.P.: Effect of geometry on the effective moisture transfer diffusion coefficient. J. Food Eng. **30**, 433–447 (1996)
132. Upadhyay, A., Sharma, H.K., Sarkar, B.C.: Characterization and dehydration kinetics of carrot pomace. Agricultural Engineering International: The CIGR Ejournal. Manuscript FP 07 35. **10**(35) (2008)
133. Van den Berg, C., Bruin, S.: Water activity and its estimation in food systems. In Rockland, L.B., Stewart, G.F. (eds.), Water Activity: Influences on Food Quality, pp. 147–177. Academic Press, New York (1981)
134. Vázquez, G., Chenlo, F., Moreira, R., Castoyas, A.: The dehydration of garlic. I. Desorption isotherms and modeling of drying kinetics. Dry. Technol. **17**(6), 1095–1108 (1999)
135. Wadsworth, J.I., Velupillai, L., Verma, L.R.: Microwave-vacuum drying of parboiled rice. Tran. Am. Soc. Agric. Eng. **33**(1), 199–221 (1990)
136. Walde, S.G., Balaswamy, K., Velu, V., Rao, D.G.: Microwave drying and grinding characteristics of wheat (Tricitium aestivum). J. Food Eng. **55**(3), 271–276 (2002)
137. Wang, C.Y., Singh, R.P.: A Single Layer Drying Equation for RoughRice. ASAE Press, St. Joseph (1978)
138. Wang, J., Xi, Y.S.: Drying characteristics and drying quality of carrot using a two-stage microwave process. J. Food Eng. **68**(4), 505–511 (2005)
139. Wang, W., Chen, G.: Theoretical study on microwave freeze-drying of an aqueous pharmaceutical excipient with the aid of dielectric material. Dry. Technol. **23**(9–11), 2147–2168 (2005)
140. Wang, W., Thorat, B.H., Chen, G., Mujumdar, A.S.: Simulation of fluidized-bed drying of carrot with microwave heating. Dry. Technol. **20**(9), 1855–1867 (2002)
141. Wang, Z., Sun, J., Chen, F., Liao, X., Hu, X.: Mathematical modeling on thin layer microwave drying of apple pomace with and without hot-air pre drying. J. Food Eng. **80**(2), 536–544 (2007)
142. Wang, Z., Sun, J., Liao, X., Chen, F., Zhao, G., Wu, J., Hu, X.: Mathematical modeling on hot air drying of thin layer apple pomace. Food Res. Int. **40**, 39–46 (2007)
143. White, G.M., Ross, I.J., Poneleit, C.G.: Fully exposed drying of popcorn. Trans. ASAE **24**(2), 466–468 (1981)
144. Xu, Y., Min, Z., Mujumdar, A.S., Zhou, L.-Q., Sun, J.-C.: Studies on hot air and microwave vacuum drying of wild cabbage. Dry. Technol. **22**(9), 2201–2209 (2004)
145. Yadav, A.K., Singh, S.V.: Osmotic dehydration of fruits and vegetables: a review. J. Food Sci. Technol. (2012) doi:10.1007/s13197-012-0659-2, published online 22 February 2012
146. Yaldiz, O., Ertekyn, C.: Thin layer solar drying of some vegetables. Dry. Technol. **19**(3–4), 583–597 (2001)
147. Yaldiz, O., Ertekin, C., Uzun, H.I.: Mathematical modeling of thin layer solar drying of sultana grapes. Energy **26**, 457–465 (2001)
148. Yongsawatdigul, J., Gunasekaran, S.: Microwave-vacuum drying of cranberries: Part I. Energy use and efficiency. J. Food Process. Preserv. **20**(2), 121–143 (1996)
149. Yongsawatdigul, J., Gunasekaran, S.: Microwave-vacuum drying of cranberries: Part II. Quality evaluation. J. Food Process. Preserv. **20**(2), 145–156 (1996)
150. Zhang, S., Zheng, B.: Effect of drying methods on quality of abalone. J. Food Agric. Environ. **11**(3–4), 444–447 (2013)

Convective Drying of Food: Foundation, Modeling and Applications

A. G. Barbosa de Lima, R. P. de Farias, W. P. da Silva, S. R. de Farias Neto, F. P. M. Farias and W. M. P. B. de Lima

Abstract This chapter focuses in a theoretical and experimental study of food dehydration with particular reference to drying of fruits and vegetables. Topics related to fundamentals, theory and effect of drying and modeling are presented and discussed. Application has been done to drying of banana fruit. Whole bananas were peeled manually and dried in an oven at temperatures ranging from 40 to 70 °C. Drying, heating and shrinkage lumped models were proposed and fitted to experimental data. Non-linear regression analyses were done to verify the consistence of the models to predict the experimental data. Results revealed which air temperature

---

A. G. Barbosa de Lima (✉) · W. M. P. B. de Lima
Department of Mechanical Engineering, Federal University of Campina Grande, Av. Aprígio Veloso, 882 Bodocongó, Campina Grande, PB 58429-900, Brazil
e-mail: gilson@dem.ufcg.edu.br

W. M. P. B. de Lima
e-mail: wan_magno@hotmail.com

R. P. de Farias
Department of Agriculture Science, State University of Paraiba, Catolé do Rocha, PB, Brazil
e-mail: rp.eng.tec@gmail.com

W. P. da Silva
Department of Physics, Federal University of Campina Grande, Av. Aprígio Veloso, 882 Bodocongó, Campina Grande, PB 58429-900, Brazil
e-mail: wiltonps@uol.com.br

S. R. de Farias Neto
Department of Chemical Engineering, Federal University of Campina Grande, Av. Aprígio Veloso, 882 Bodocongó, Campina Grande, PB 58429-900, Brazil
e-mail: fariasn@deq.ufcg.edu.br

F. P. M. Farias
Department of Technology and Development, Federal University of Campina Grande, Sumé, PB, Brazil
e-mail: fabianapimentel@ufcg.edu.br

J. M. P. Q. Delgado and A. G. Barbosa de Lima (eds.), *Transport Phenomena and Drying of Solids and Particulate Materials*, Advanced Structured Materials 48,
DOI: 10.1007/978-3-319-04054-7_5,

affect significantly moisture removal, heating and dimensions variations rates and quality of banana fruit. The fitted models presented good concordance with experimental data.

Keywords Drying · Experimental · Lumped model · Simulation

1 Basic Concept in Food

Food is any substance consumed to provide nutritional support for the body. It is ingested and assimilated by an organism to produce energy, stimulate growth, and maintain life.

Food products exist in the solid and liquid phases. They present great variability in composition and physical characteristics. For example, the composition of grains, fruits and vegetables depends on variety, site variables, climate, etc. [1].

Constituents commonly found in foods include moisture, protein, fat, carbohydrate, fiber, and ash [2]. The volatile part of a food item can be termed moisture. Moisture in the form of water molecules is bonded to various parts of the product in varying ways as follows: (a) ionic groups, such as carboxyl and amino acids, and (b) hydrogen groups, such as hydroxyl and amides. In high-moisture foods in which moisture contents are more than 50 % wet basis, unbound free water exists in the interstitial pores and in intercellular spaces [3].

Because water (continuous liquid phase) is the predominant constituent (highest concentration) in most fresh foods, water concentration significantly influences the palatability, digestibility, physical structure, technical handling ability, and thermophysical properties of food materials [2, 4].

For agricultural and horticultural products, water content varies with the cultivar as well as with the stage of development or maturity when harvested, the growing conditions, and amount of moisture lost after harvest [2].

Most of the deteriorative processes that take place in foodstuffs are influenced by the temperature, and concentration and mobility of water inside the food. Perishable food materials are more stable at low than at high moisture content, and low storage temperature [4]. Thus, one of the prime goals of food processing on preservation is to convert perishable foods such as grains, fruits and vegetable in stabilized products, consequently reducing postharvest losses, in order to extend storage self life and to quality enhancement.

Today, several process technologies have been employed on an industrial scale to preserve fresh foods, such as packing, canning, cooling, freezing, heating, and dehydration [5].

A food product is in equilibrium with its surrounding when its internal vapor pressure becomes in equilibrium with the outside vapor pressure. This condition occurs as the wet material has been exposed to a particular air condition for a sufficiently long period of time (steady-state limit condition). The moisture content of the product at this stage is called the equilibrium moisture content [3] and its magnitude is a function of the structure and type of the subject food and of the prevailing drying condition [6]. It has been used as reference to safe storage of the perishable materials.

Thus, there is a wide range of techniques available in the literature for measuring the wet material moisture content. They are divided in direct and indirect methods. In direct methods the following can be cited: material heating, chemical drying, azeotropic distillation, chemical methods, gas chromatography and refractometry. In indirect methods the following can be distinguished: electrical conductivity, capacitance, microwaves, infrared absorption, equilibrium relative humidity and temperature difference [7, 8].

2 Food Dehydration Foundations

2.1 Fundamentals

Food dehydration is one of energy-intensive unit operations that has been used to removal the moisture from solids food as an integral part of food processing. Dehydration can be realized by heating (drying) and osmotic pressure difference (osmotic dehydration).

Dehydration of fresh foods generally involves a series of interdependent unit operations like blanching, pasteurization, preconcentration, and drying, all of which contribute to the overall quality of the final product after processing. Dehydration by drying involves simultaneous transfer of heat, mass and momentum in which heat penetrates into the product and moisture is removed by evaporation into an unsaturated gas phase.

Many perishable food products of great nutritional importance for peoples, such as fruits and vegetables, are dehydrated in the food industry. The loss of nutrients and functional attributes varies with regard to the type, condition and time of dehydration process and the sensitivity of specific food components. During dehydration the product is generally above room temperature but well below sterilization temperature. The added heat and exposure time of the product at elevated temperature affect the nutrient quality of the food products, especially that very sensitive to heating.

The different types of food degradation during dehydration by drying are chemical (enzymatic and non enzymatic browning reaction, lipid oxidation, color loss), physical and structural (rehydration, solubility, texture, and aroma loss) and nutritional (vitamin loss, protein loss, and microbial changes). All changes depend on water activity, temperature and exposure time [3, 6].

The aim of dehydration is to prevent the growth and reproduction of microorganisms causing decay and minimizes many of the moisture-mediated deteriorative reactions. It brings about substantial reduction in weight and volume, minimizing packing, storage, and transportation costs and enables storability of the product under ambient temperature [5]. Thus, following advantages of this thermal process can be cited [3]:

a. Extended storage life;
b. Quality enhancement;
c. Ease of packaging, handling, and transportation;
d. Sanitation (Destruction of insects and the microorganisms);
e. Improved milling, mixing, or segregations.

Most perishable food products, such as, fruits and vegetables contain more than 80 % water and they are, therefore, highly perishable. Water loss and decay account for most of their loss, which are estimated to be more than 30–40 % in the developing countries in the tropics and subtropics due to inadequate handling, transportation, and storage facilities.

Consequently, the need to reduce postharvest losses of these perishable horticultural foods is of paramount importance for developing countries, in order, to increase their availability, especially in the present context when the constraints on food production are continually increasing [5].

Heat and mass transfer plays a very important role in different unit operations of food processing, such as drying, extraction, distillation, and absorption [9]. They involve moisture and heat fluxes in solid foods and interphase heat and moisture transfer in food processing and storage.

Moisture flux within the solid porous food during drying, rehydration, or storage is a complex physical process that depending on the size, shape, and connectivity of the pores may take place via mechanisms molecular diffusion, capillary flow, Knudsen flow, hydrodynamic flow, or surface diffusion. Molecular diffusion is the mass transfer provoked by random movement of molecules. Knudsen diffusion occurs when the mean free path of the diffusing molecules is large in comparison with the diameter of the capillary [9, 10].

Mass diffusion of gases, vapors, and liquids in solids media is more complex process than diffusion in liquids. The porous solids usually have a heterogeneous structure, and they may interact with the diffusing compounds. Consequently, the diffusion velocity of small molecules in these solids is much lower than in liquids, and this may affect the various physical and chemical processes velocities involving mass transfer [9].

Diffusion in gases can be treated in terms of kinetic theory, but an empirical treatment is necessary for diffusion in solids. In this context, mass diffusivity is a physical property of the system (diffusing substance and medium) that represents as quickly the moisture migrates within the porous solid [9].

In isotropic porous solids, diffusion can be expressed in terms of an effective mass diffusivity (D_e), which is smaller than the molecular diffusivity (D_{AB}), being defined as follows [11]:

$$D_e = \frac{\varepsilon D_{AB}}{\tau} \quad (1)$$

where ε is porosity (void fraction) of the solid and τ is the tortuosity, a factor that connects for the long tortuous path through the pores (varying from 1.5 to 5) [9].

Estimation of the effective diffusion coefficient in isotropic macroporous media, has also been proposed by van Brakel and Heertjes [12] as follows:

$$D_e = \left(\frac{\delta\varepsilon}{\tau^2}\right) D_{AB} \tag{2}$$

In this equation δ represent the constrictivity, and D_{AB} is the vapor diffusivity in air in the absence of porous media [13]. The constrictivity accounts for the fact that the cross section of a pore space segment varies over its length. Many fresh foods present a porous structure, which is developed during the drying process, especially in process at high temperature, when water is removed as a vapor. The porous structure can be characterized by the bulk porosity or void fraction of the material, which can be estimated by using the following equation [14]:

$$\varepsilon = 1 - \frac{\rho_b}{\rho_p} \tag{3}$$

where ρ_b is the bulk density and ρ_p the particle (solid) density.

Effective or apparent moisture diffusivities, reported in the literature, have been estimated, usually from drying or sorption rate experimental data at a specified temperature [9].

The diffusivity of the various compounds in the porous solid depends on the temperature, and the following form of the Arrhenius equation has been applied:

$$\frac{d\ \ln D_e}{dT} = -\frac{E}{RT^2} \tag{4}$$

where E is the energy of activation for diffusion, which may vary with the concentration of the diffusant in the solids [9]. Further, it is found in the literature different equations for mass diffusivity as a function of the moisture content and/or temperature [15]. In general comparison between diffusivities reported in the literature is difficult because of the different methods of estimation and the variation of food composition and physical structure. The need for more reliable data of diffusivity is obvious [9].

In porous (granular extruded, puffed) foods, the effective moisture diffusivity increases gradually as the moisture content is reduced to about 0.1, evidently due to the development of porosity [9].

The diffusion of volatile aroma compounds is very important in food processing and storage, and an understanding of the mechanism of mass transfer will help in maintaining the food quality. On the basis of high relative volatility, high losses of volatiles would be expected during evaporation and drying of food products. However, those components may be retained at a relatively high percentage in the dried product because of the presence of soluble and insoluble solids in the food. Retention of aroma compounds depends on the concentration and nature of the food solids. During the fast evaporation in liquids food or wet solids foods, thermodynamic equilibrium is not reached with the very volatile components [9].

Interphase mass transfer plays an important role in different processes such as evaporation, drying, freezing, and storage of foods. During drying of wet porous materials simultaneous heat and mass transfer occurs both inside and in the boundary layer of the drying agent surrounding it. In general, in the drying process a considerable influence is exerted both by the external conditions and by the internal structure of the material to be dried. In this context, hygroscopic porous solids are more likely to exhibit internal resistance to moisture migration for some materials, particularly foodstuffs, because their internal cellular structure and the presence of shell. The relative significance of the internal and external heat and mass transfer process can be expressed by Kirpichev number or Biot number [7]. If the resistance to mass transfer at the surface of the porous solid is significant compared to the resistance within the solid, it must be considered. For this purpose the Biot number of mass transfer has been used. This parameter is defined by the following equation:

$$Bi = \frac{hL}{D_e} \tag{5}$$

where h is the convective mass transfer coefficient at the surface of the solid. By analyzing the Eq. (5), we notice that high Biot number can be obtained by increasing the mass transfer coefficient [9].

Similar equation for Biot number has been applied to heat transfer process by changing diffusion coefficient by thermal conductivity and convective mass transfer coefficient by convective heat transfer coefficient.

The literature contains very few experimental data on mass (or heat) transfer coefficients pertinent to food systems.

Interphase mass (or heat) transfer rates can be increased by increasing the air—velocity and /or temperature. Centrifugal force may increase the mass transfer rate, and a centrifugal fluidized bed has been proposed to accelerate air—drying of fruits and vegetables [9, 16].

2.2 Prediction in Drying

In recent days, with the drying of food materials the following trends are found [10]:

(a) Understanding of the physical and mathematical aspects of the dehydration process;
(b) Reduction of the changes in foodstuffs during dehydration;
(c) Optimization of the drying process.

Modeling of the drying process ultimately has presented great potential to quickly identify the consequences of schedule alterations and provide focus to the solution of drying problems. The modeling of wet porous solids drying includes two inter-related areas [17]: (a) modeling of single-particle drying or collection of particle (thin-layer model) and (b) modeling of deep-bed layer drying (dryer model).

Mathematical modeling and numerical simulation not only allows some rather expensive and repetitive experimentation to be avoided, but perhaps most importantly, can be used to elucidate on the underlying physic associated with the convoluted heat and mass transport phenomena which are generated within the porous media during the drying process [18].

It is very difficult to mathematically describe the transport of water in structured food material. Despite of this situation, several empirical, semi-empirical, and theoretical models have been proposed to describe drying process. Theoretical models take into account only internal resistance while the semi-empirical and empirical models (thin-layer drying models) take into account only external resistance to heat and moisture transfer between wet porous solid and air (internal resistance is negligible). Further, as shrinkage in fruits is an observable phenomenon and may have significant effect on mass diffusivity, it must be taken into account in order to obtain reliable predictions of performance. The changes in the volume of a solid during diffusion can be taken into consideration by using suitable frames of reference based on the dry solids content of the material.

Because the simplifications, the empirical and semi-empirical models cannot give clear and accurate of the fundamentals of the drying process, but they can used to complete mathematical models at the level of the drying equipment or supply information of the average moisture content and temperature at a specified drying time.

In general, the non-steady state mass diffusion in solids can be treated mathematically in manner similar to the heat conduction (Fourier's law).

The moisture transfer in heterogeneous porous media can be conveniently analyzed by using Fick's second law applied to homogeneous porous materials, in which the heterogeneity of the material is accounted by using an effective diffusivity.

Moisture diffusivity depends strongly on temperature and, often, very strongly on the moisture content. Further, as reported before, the void fraction affects diffusivity significantly, and the pore structure and distribution do so even more [13].

In foods, mass diffusion is often moisture concentration dependent. Because of this non-linearity the differential equation of diffusion must be solved by numerical or other special methods. On the other hand, when strong interactions between the diffusing component and the porous medium occur, for example, swelling, diffusion does not obey Fick's law (non-Fickian diffusion). Non-Fickian diffusion or anomalous diffusion may be directly related to the influence of the changing structure on solubility and diffusional mobility, or they may result from the internal stresses exerted by one part of the porous media an another as diffusion proceeds [19].

Exact solution of the non-steady state (transient) diffusion equation for constant mass diffusivity (or thermal diffusivity) and various boundary conditions are available in the literature for the basic shapes such as slab, finite and infinite cylinder, parallelepiped, and sphere [19–23].

The mass diffusivity of various substances in solid media is of special interest to food engineering, because most mass transport operations in the food industry involve solid or semisolid foods. Literature data are very limited, and they vary considerable because of the complex structure of foods and the lack of a standard method for determination of diffusivity [9].

The mechanism of water transport may change from liquid diffusion at high moisture to vapor diffusion, which is much faster. At low moisture (<0.1) the mass diffusivity drops sharply because of the difficulty of removing the strongly sorbed water molecules from solid matrix [9, 24].

Now that we have discussed about moisture diffusion in wet food materials to be dried, the next step is to perform application related to this topic.

3 Application: Drying of Banana Fruit

3.1 Background

Several types of dryers and drying methods, each better suited for a particular situation, are commercially used to remove moisture from a wide variety of fruits and vegetables. Factors on which the selection of a particular dryer/drying method depends include form of raw material and its properties, desired physical form and characteristics of the product, necessary operating conditions, and operation costs. In general, the following quality changes during drying and storage of fruits can be cited [5]:

(a) Loss of vitamins (vitamins A and C);
(b) Loss of natural pigments (carotenoids and chlorophylis);
(c) Discoloration due to enzymatic or nonenzymatic browning;
(d) Oxidative degradation and flavor loss;
(e) Irreversible damage to the texture (shrinkage, slow cooking, and incomplete rehydration);
(f) Loss of water reducing microbiological spoilage.

Fruits and vegetables play an important role in human diet and nutrition as sources of vitamins and minerals. Overall post-harvest losses of fruit and vegetables in developing countries are estimated at about 20–50 % of the production. Fresh fruits are dried after harvesting, in order to reduce the waste and the spoilage, and to extend their shelf life. During the drying of fruits and vegetables occurs big variations of the volume and surface area of the product simultaneously with the loss of moisture and increase of the temperature. Therefore, it is necessary to devote more attention to shrinkage phenomenon, because it affects the drying and heating rates.

Bananas are fruits with aromatic flavor which are naturally sweet. They contain different constituents such as fat, high natural sugars content, protein, potassium and vitamins A, B complex and C. They provide flavor and variety to the human diet, and also serve as an important source of essential vitamins and minerals, although they are not good and economical sources of protein, fat and energy. From the viewpoint of production and consumption banana fruit post-harvested contains high moisture content, and therefore very perishable and susceptible to fast deterioration during transportation and storage. This in turn causes serious economic losses as a result of reduction in weight and quality.

Because the sequential growth of banana cultures and the amount of harvested banana, losses of these fruits has increased and thus generates the necessity of studies for conservation of these fruits [25–27]. Commercially, banana, whole or in slices, is dried from approximately 80 % to less than 20 % final moisture content [28, 29] or down to 14–15 % final moisture content (on dry basis) [30]. Due to the high sugar and water contents, bananas are dried normally in high temperatures and prolonged drying times, which can cause serious adverse changes in the finished products [31].

In order to accurately understand moisture migration phenomenon within biological products, mainly foodstuffs (grains, fruits, vegetables, etc.), and to explain the effects of drying on the quality of the material, the moisture and heat transfer in individual particles should be understood and accurately represented by a mathematical model. Now, this chapter presents both theoretical (drying, heating and shrinkage lumped models) and experimental drying study of banana fruit.

3.2 Experimental Procedure

Farias [32] and Farias et al. [33] conducted several batch drying experiments of peeled whole banana. Ripened banana fruits (Musa species and "prata" variety), purchased from a local super market were used in the investigation, immediately. The whole fruits were hand peeled. Measurements of the thickness, length and diameter were made at different points with a digital paquimeter and initial weight was measured using a digital electronic balance. All bananas used for drying were from the same batch. Drying experiments were performed for peeled whole banana using a laboratory-scale system (oven) in different drying temperatures. The air velocity was set at a constant 1 m/s. Fruit was placed into the oven on a stainless steel mesh tray. Figure 1 shows one the fruits used in the drying experiments. Table 1 shows the experimental information about the fruit and drying air.

Moisture loss was measured by periodically taking out and weighting the banana using a digital electronic balance with 0.1 g precision; surface temperature was measured using an infrared thermosensor. Following, we putting the sample within an oven for continue drying until the next time point. Each time interval for the measurement was 5 min. The drying continued until there was no significant decrease of the fruit moisture with increasing the drying time between consecutive measurements. After this instant of drying the fruit was maintained for 24 h at the same temperature, in order to obtain the moisture in the equilibrium condition.

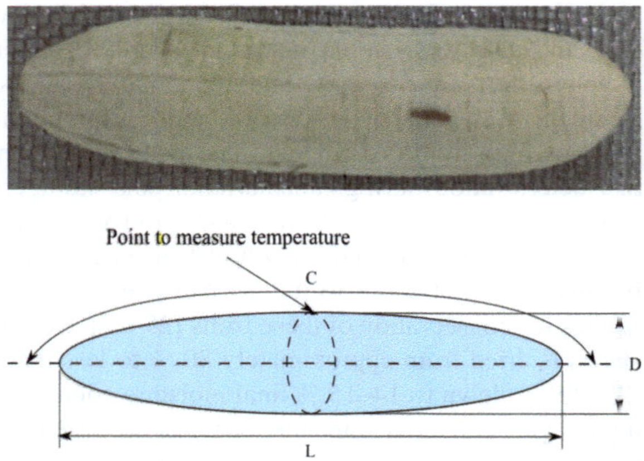

Fig. 1 Banana fruit and geometrical representation

Table 1 Experimental parameters of the air and whole banana used in this work

Air					Whole banana			
T (°C)	RH (%)	v (m/s)	C (mm)	L (mm)	D (mm)	M_o (d.b.)	M_e (d.b.)	θ_o (°C)
40	29.58	0.04	116	112.76	29.65	2.0534	0.1441	26.7
50	17.32	0.05	110	97.60	26.72	2.3669	0.0672	27.2
60	10.15	0.06	136	122.39	31.42	2.3752	0.0452	27.3
70	6.93	0.07	115	105.36	27.83	2.5541	0.0015	27.6

This moisture was used to calculate the value of equilibrium moisture content. Following, the fruit was maintained into the oven for more 24 h at temperature of 70 °C, in order, for obtain dry matter mass.

3.3 Mathematical Treatment

In order to describe the drying behavior of banana, and predict it under different drying conditions, it is necessary to model the drying process. Drying of banana predominantly follows a falling rate profile. Because the high moisture content of fruit post-harvested mass transfer during this period is mainly caused by liquid diffusion. As results, the rate of diffusion is governed by moisture concentration gradient as the driving force. Thus, Fick's law of diffusion and empirical models can be used to model the drying behavior for this period. At present, there are very few models that represent the batch drying of tropical fruits specially banana. Herein, it was propose models to moisture removal, heating and shrinkage of the banana fruit.

(a) Drying model

Mass diffusion equation as applied to simple geometry, for example, plate, cylinder, sphere and ellipsoid, has exact solution which average value of the moisture content is given for a convergent infinite series as follows [19–23, 34, 35]:

$$(\overline{M} - M_e)/(M_o - M_e) = \sum_{n=1}^{\infty} A_n \exp(-K_n t) \tag{6}$$

where Na and K_n are dependent of the body shape and specified boundary condition.

In the Eq. (6), the successive terms of the infinite series decrease by increasing the numbers of terms (n). For long drying time, convergence must be faster, and few terms of the series are sufficient to calculate the average value of the moisture content. Thus, considering n = 3, Eq. (6) assumes the form presented in Eq. (7), as follows:

$$(\overline{M} - M_e)/(M_o - M_e) = A_1 \exp(-K_1 t) + A_2 \exp(-K_2 t) + A_3 \exp(-K_3 t) \tag{7}$$

(b) Heating model

The model to predict surface temperature of fruit during drying was based on the works of Azzouz et al. [36], Pérez [37] and Lima et al. [38]. It is given by Eq. (8) following:

$$(\theta - \theta_e)/(\theta_o - \theta_e) = A_1 + A_2 \log_{10}\left(t^{K_1} + A_3\right) \tag{8}$$

(c) Shrinkage model

During drying process, the porous body changes dimensions by heating (dilation) and moisture removal (contraction, well-known as shrinkage). Dilation phenomenon is insignificant as compared with shrinkage, because the volumetric dilation coefficient is smallest that the shrinkage coefficient, mainly for fruits and vegetables. However, when moisture movement from the body to surrounding is stopped, dimension variation due to heating is a very important phenomenon; it provokes large thermal stress within the solid, mainly for more rigid materials (post-drying). A fundamental point in the theory of shrinkage phenomenon is to explain by an equation like both volume and average moisture content are related. According by Lima [38] and Lima et al. [39, 40, 41], the following mathematical relation for linear shrinkage was proposed:

$$\frac{(V)_t}{V_o} = 1 - K_1(\bar{M}_o - \bar{M}) \tag{9}$$

In the Eqs. (7), (8) and (9), the parameters A_i and K_i are obtained by fitting the models to experimental data.

During drying process banana shrinks, modify its shape, decreases the area of heat transfer and increases the superficial roughness. This last characteristic provides an increase of the turbulence level in the boundary layer, thus, favoring transfer of energy between air and fruit. For determination of the volume [42] and surface area [43] of the banana, which was considered as a prolate spheroid it was used the following equations:

$$(V)_t = \frac{4}{3}\pi\left(\frac{L}{2}\right)_t\left(\frac{D}{2}\right)_t^2 \tag{10}$$

$$(S)_t = \frac{1}{2}\pi(D)_t(L)_t\left\{\frac{(D)_t}{(L)_t} + \frac{\arcsin\left[\sqrt{\left[1 - \left(\frac{(D)_t}{(L)_t}\right)^2\right]}\right]}{\sqrt{\left[1 - \left(\frac{(D)_t}{(L)_t}\right)^2\right]}}\right\} \tag{11}$$

3.4 Results Analysis

3.4.1 Drying Experimentation

Based on the current experimental drying characteristics data, it was plotted in Figs. 2, 3, 4, 5, 6 and 7 moisture content, surface temperature and volume variations (shrinkage) as a function of the drying time. From the analyses of these figures we can see that moisture content decreases with the time and that the drying and shrinkage rates are higher as higher temperature and lower relative humidity of the drying-air are used.

The effect of drying-air temperature on drying time showed that increase in air temperature resulted in decrease in drying time. Further, it can be seen that drying rate decreased continuously with decreasing moisture content or increasing drying time. An almost linear behavior of the volume and surface area with the moisture content has been verified. Surface temperature of the fruit varied during drying process and reached thermal equilibrium more quickly in the lower drying temperature. Thus, we notice that the drying occurred in falling drying period, for a fixed temperature, in accordance to literature [25, 31, 44–50].

From the analysis of the Fig. 7 it can be seen that temperature has different behavior. This is an indicative that both thermal diffusivity and convective heat

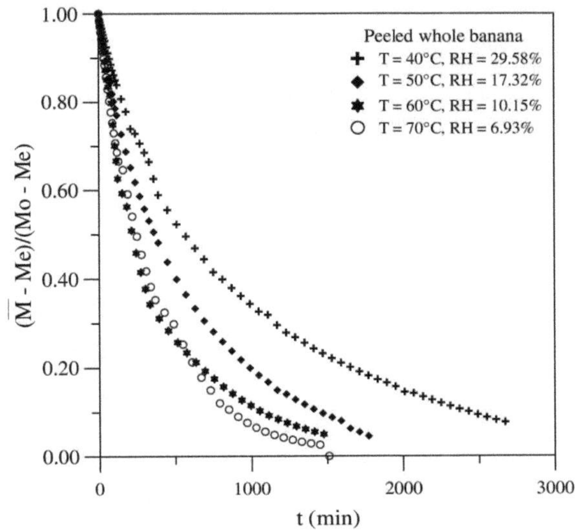

Fig. 2 Experimental dimensionless average moisture content of the whole banana as a function of drying time at different drying conditions

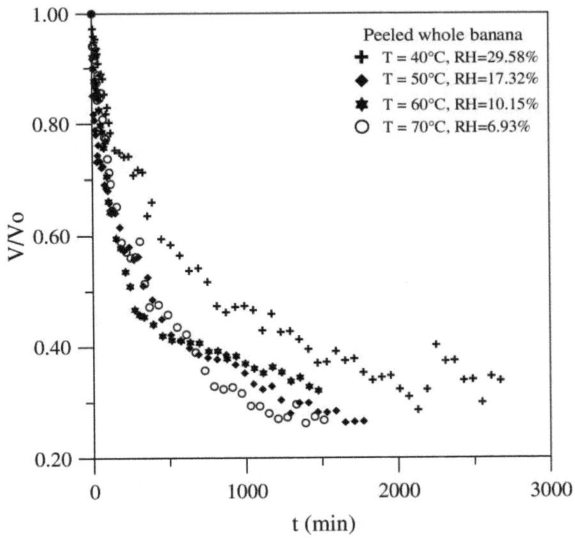

Fig. 3 Experimental dimensionless volume of the whole banana as a function of drying time at different drying conditions

transfer coefficient at the surface of fruit are affected by volume and temperature variations.

Table 2 presents the dimensionless length, dimensionless surface area and volume for each experimental test. From the analysis of this table we notice that the shrinkage of the fruit correspond almost to the volume of water removed during drying for each experiments. The, according to Eq. (9), there is almost linearity between the volume variation and moisture content. Shrinkage is more

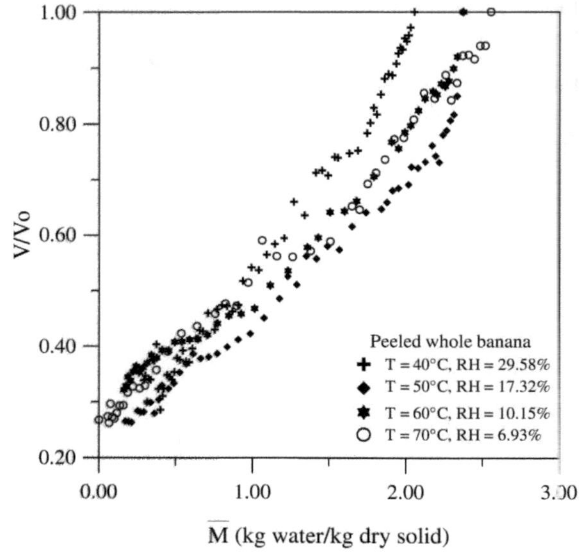

Fig. 4 Experimental dimensionless volume of the whole banana as a function of average moisture content at different drying conditions

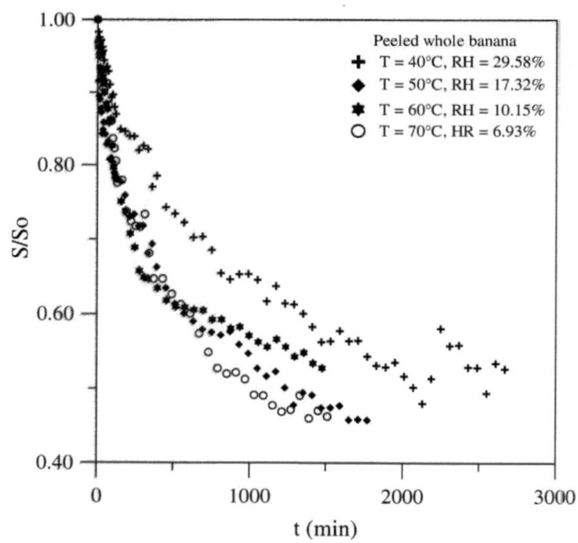

Fig. 5 Experimental dimensionless surface area of the whole banana as a function of drying time at different drying conditions

intense in higher air temperature. Similar behavior occurs with surface area. Dimensionless data of the diameter and length are different, so, we can say that the shrinkage velocities are different in both radial and axial directions of the fruit. Thus, it was verified a non-uniform shrinkage. Area/volume relationships remain approximately constant for all samples.

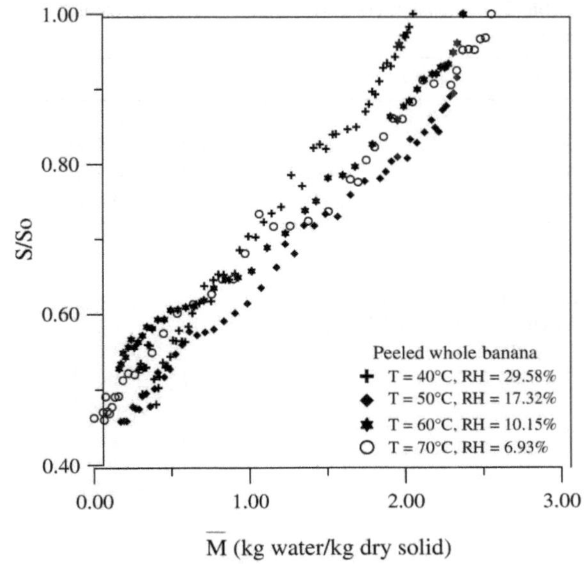

Fig. 6 Experimental dimensionless surface area of the whole banana as a function of average moisture content at different drying conditions

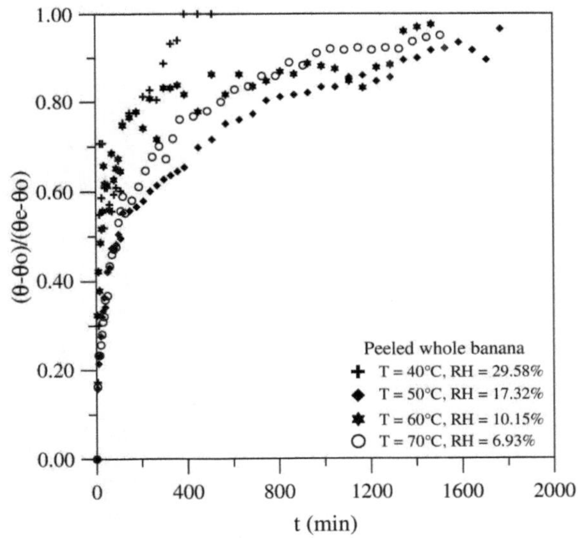

Fig. 7 Experimental dimensionless surface temperature of the banana as a function of drying time at different drying conditions (see Fig. 1)

Figure 8 illustrates the fruit at different instants of drying at temperature 70 °C. We can see that the surface hardness, shape variation and loss of color were verified in the fruit post-drying.

Table 2 Dimensionless geometrical data of the whole banana at the initial and final stages of drying

T (°C)	t (min)	D_f/D_o	L_f/L_o	S_f/S_o	V_f/V_o	M_o (w.b.)	S_o/V_o
40	2670	0.6361	0.8374	0.5269	0.3388	0.6725	0.1634
50	1770	0.5711	0.8135	0.4580	0.2634	0.7030	0.1817
60	1475	0.6002	0.8916	0.5279	0.3213	0.7039	0.1540
70	1450	0.5738	0.8313	0.4705	0.2737	0.7186	0.1741

Fig. 8 Banana fruit at different moments of drying (T = 70 °C)

Table 3 Parameters of the drying equation obtained by fitting to experimental data

T (°C)	Parameters						R^2 (%)	Loss function $(obs\text{-}pred)^2$
	A_1	K_1	A_2	K_2	A_3	K_3		
70	−0.030115	0.000232	0.15645	0.009797	0.879095	0.002182	99.941	0.0032339

Table 4 Parameters of the heating equation obtained by fitting to experimental data

T (°C)	Parameters				R^2 (%)	Loss function $(obs\text{-}pred)^2$
	A_1	A_2	A_3	K_1		
70	−0.269941	0.364485	6.298940	1.066960	99.425	0.01707067

Table 5 Parameters of the shrinkage equation obtained by fitting to experimental data

T (°C)	Parameter	R^2 (%)	Loss function $(obs\text{-}pred)^2$
	K_1		
70	0.300195	96.906	0.81233047

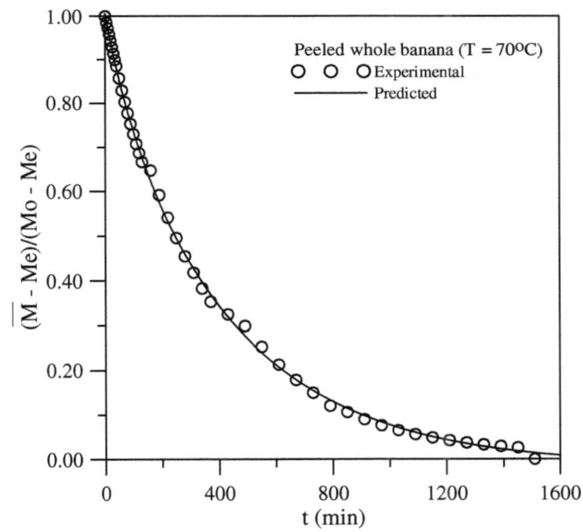

Fig. 9 Dimensionless average moisture content of the banana during drying at 70 °C

3.4.2 Drying Simulations

The constants and coefficients of the thin layer-models as applied to drying at 70°C were computed through non-linear regression analysis. The various statistical parameters corresponding to the models (coefficient of determination and variance) were determined using the Hooke-Jeeves numerical method with convergence criterion 0.0001. All details of the results and statistical parameters are presented in Tables 3, 4 and 5 and Figs. 9, 10 and 11.

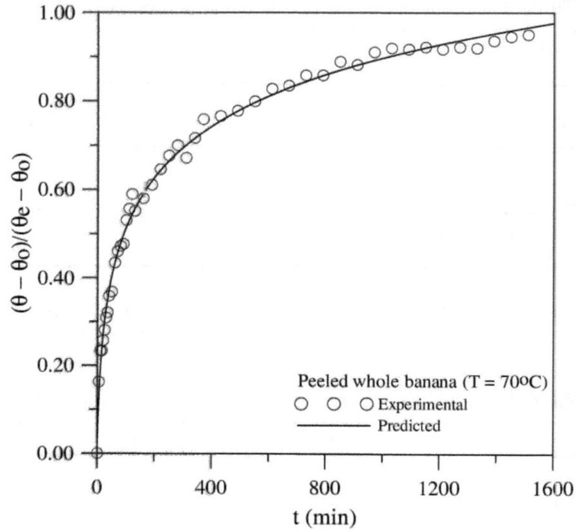

Fig. 10 Dimensionless surface temperature at the surface of the banana during drying at 70 °C (see Fig. 1)

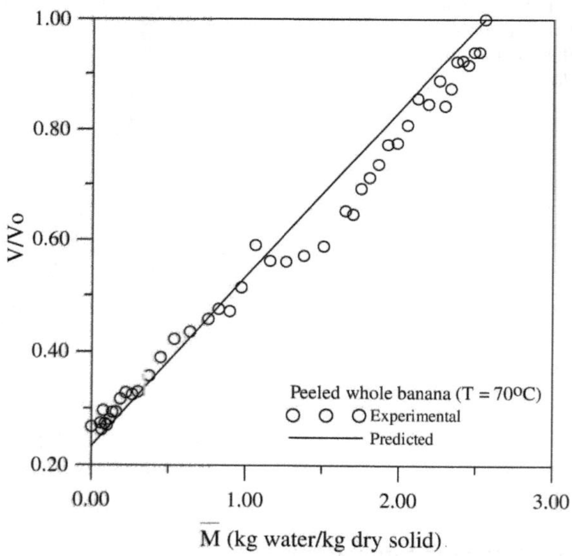

Fig. 11 Dimensionless volume of the banana during drying at 70 °C

The statistical parameters were used to determine the consistency of the models for the current experimental conditions. The values for the coefficient of determination (R^2) were greater around 0.999 for moisture content, 0.994 for temperature and 0.969 for shrinkage, as a result, the proposed models may be assumed to represent the drying behavior of whole banana.

4 Concluding Remarks

In this chapter transport phenomena (heat conduction, mass diffusion and dimensions variations) in wet porous solid has been explored, with special reference to dehydration by drying. Interest in this class of physical problem is motivated by its importance in many industrial applications (for example, food and chemical processing). Here, our attention is focused on drying of banana fruit drying.

The drying, heating and shrinkage kinetics of banana were analyzed with one drying technique (convection in oven). For unsaturated air condition, the analysis showed that air temperature has greatly affected moisture removal, temperature and dimensions of the fruit. Lumped models (drying, heating and volume variation) available in the literature were fitted for all the drying experimental conditions and the coefficients of the models were established through the non-linear regression analysis. The applicability of the models was examined through the statistical parameters and was found that all the predicted data agreed well with experimental data. Finally, we would like to cite that the proposed models with the new constants can be satisfactorily used to understand the drying characteristics and to aid engineers in optimized design of the banana processing units.

Acknowledgements The authors would like to express their thanks to CNPq (Conselho Nacional de Desenvolvimento Científico e Tecnológico, Brazil), CAPES (Coordenação de Aperfeiçoamento de Pessoal de Nível Superior, Brazil), and FINEP (Financiadora de Estudos e Projetos, Brazil) for supporting this work; to the authors of the references in this paper that helped in our understanding of this complex subject, and to the Editors by the opportunity given to present our research in this book.

References

1. Sweat, V.E.: Thermal properties of foods. In: Rao, M.A., Rizvi, S.S.H. (eds.) Engineering properties of foods, pp. 99–138. Marcel Dekker Inc, New York (1995)
2. ASHRAE: Thermal properties of foods. In: ASHRAE Refrigeration Handbook. American Society of Heating, Refrigeration and Air-Conditioning Engineers, Inc., Atlanta. p. 8.1 (1998)
3. Sokhansanj, S.: Drying of foodstuffs. In: Mujumdar, A.S. (ed.) Handbook of industrial drying, vol. 1, pp. 589–625. Marcel Dekker Inc., New York (1995)
4. Wolf, W., Spiess, W.E.L., Jung, G.: Sorption isotherms and water activity of food materials. Elsevier, New York (1985)
5. Jayaraman, K.S., Das Gupta, D.K.: Drying of fruits and vegetables. In: Mujumdar, A.S. (ed.) Handbook of Industrial drying, vol. 1, pp. 643–690, Marcel Dekker, Inc., New York (1995)
6. Rizvi, S.S.H.: Thermodynamic properties of foods in dehydration, In: Rao M.A., Rizvi, S.S.H. (eds.) Engineering properties of foods. pp. 223–309. Marcel Dekker., New York (1995)
7. Strumillo, C., Kudra, T.: Drying: principles, science and design. Gordon and Breach Science Publishers, New York (1986)
8. Molnár, K.: Experimental techniques in drying. In: Mujumdar, A.S. (ed.) Handbook of Industrial drying, vol. 1, pp. 41–70. Marcel Dekker Inc., New York (1995)

9. Saravacos, G.D.: Mass transfer properties of foods. In: Rao, M.A., Rizvi, S.S.H. (eds.) Engineering properties of foods, pp. 169–221. Marcel Dekker Inc., New York (1995)
10. Bruin, S., Luyben, KChAM: Drying of food materials: a review of recent developments. In: Mujumdar, A.S. (ed.) Advances in drying, vol. 1, pp. 155–215. Hemisphere Publishing Corporation, Washington (1980)
11. Geankoplis, C.J.: Transport processes and unit operations, 2nd edn. Allyn and Bacon, Boston (1983)
12. van Brakel, J., Heertjes, P.M.: Analysis of diffusion in macroporous media in terms of a porosity, a tortuosity and a constrictivity factor. Int. J. Heat Mass Transfer. **17**, 1093–1103 (1974)
13. Marinos-Kouris, D., Maroulis, Z.B.: Transport properties in the drying of solids. In: In: Mujumdar, A. S. (ed.) Handbook of Industrial drying, vol. 1, pp. 113–159. Marcel Dekker, Inc., New York (1995)
14. Marousis, S.N., Saravacos, G.D.: Density and porosity in drying starch materials. J. Food Sci. **5**, 1367–1372 (1990)
15. Zogzas, N.P., Maroulis, Z.B., Marinos-Kouris, D.: Moisture diffusivity data compilation in foodstuffs. Drying Technol. **14**(10), 2225–2253 (1996)
16. Lazar, M.E., Farkas, D.F.: The centrifugal fluidized bed. 2. Drying studies on piece-form foods. J. Food Sci. **36**(2), 315–319 (1971)
17. Pang, S., Haslett, A.N.: High–temperature kiln drying of softwood timber: the role of mathematical modeling. In: Turner, I., Mujumdar, A.S. (eds.) Mathematical modeling and numerical technique in drying technology, pp. 179–219. Marcel Dekker Inc., New York (1997)
18. Turner, I., Perré, P.: A synopsis of the strategies and efficient resolution techniques used for modeling and numerically simulation the drying process. In: Turner, I., Mujumdar, A S. (eds.) Mathematical modeling and numerical technique in drying technology, pp. 1–81. Marcel Dekker Inc., New York (1997)
19. Crank, J.: The mathematics of diffusion. Oxford Science Publications, New York (1992)
20. Carslaw, H.S., Jaeger, J.C.: Conduction of heat in solids. University Press, New York (1959)
21. Gebhart, B.: Heat conduction and mass diffusion. McGraw-Hill Inc., New York (1993)
22. Luikov, A.V.: Analytical heat diffusion theory. Academic Press Inc., Ltd, London (1968)
23. Incropera, F.P., DeWitt, D.P.: Fundamentals of heat and mass transfer. John Wiley & Sons, New York (2002)
24. Leslie, R.B., Carrillo, P.J., Chung, T.Y., Gilbert, S.G., Hayakawa, K., Marousis, S., Saravacos, G.D., Solberg, M.: Water diffusivity in starch-based systems. In: Levin, H., Slade, L. (eds.) Water relationships in foods, pp. 365–390. Plenum, New York (1991)
25. Karim, A.M.D., Hawlader, M.N.A.: Drying characteristics of banana: theoretical modeling and experimental validation. J. Food Eng. **70**, 35–45 (2005)
26. Mariani, V.C., Lima, A.G.B., Coelho, L.S.: Apparent thermal diffusivity estimation of the banana during drying using inverse method. J. Food Eng. **85**, 569–579 (2008)
27. Nguyen, M.H., Price, W.E.J.: Air-drying of banana: Influence of experimental parameters, slab thickness, banana maturity and harvesting season. J. Food Eng. **79**(1), 200–207 (2007)
28. Bowrey, R.G., Buckle, K.A., Hamey, I., Pavenayotin, P.: Use of solar energy for banana drying. Food Technol. Aust. **32**(6), 290–291 (1980)
29. Robinson, A.A.: Research design and development of banana dehydration process. Food Engineering. UNSW, Sydney, (1980)
30. Garcia, R., Leal, F., Rolz, C.: Drying of bananas using microwave and air ovens. Int. J. Food Sci. Technol. **23**(2), 73–80 (1988)
31. Maskan, M.: Microwave/air and microwave finish drying of banana. J. Food Eng. **44**, 71–78 (2000)
32. Farias, R.P.: Drying of banana in oven: thermal and geometric effects. Doctorate thesis. Process engineering, Federal University of Campina Grande. Campina Grande, Brazil (2011)
33. Farias, R.P., Silva, E.G., Lima, W.M.P.B., Silva, W.P., Lima, A.G.B. : Drying of banana: a theoretical and experimental investigation. Deff. Diff. Forum, 2014

34. Haji-Sheikh, A., Sparrow, E.M.: Transient heat conduction in a prolate spheroidal solid. Trans. ASME J. Heat Transf. **88**(3), 331–333 (1966)
35. Oliveira, V.A.B., Lima, A.G.B.: Mass diffusion inside prolate spherical solids: An analytical solution. Braz. J. Agro-ind. Prod. **4**(1), 41–50 (2002)
36. Azzouz, S., Guizani, A., Belguith, A.: Experimental analysis of heat and mass transfer during grape air drying. In: Proceedings of the 10th International Drying Symposium (IDS '96), vol. B, pp. 881–887. Krakow (1996)
37. Pérez, V.H.: Study of the behavior of temperature of banana during the drying process. Master thesis, State University of Campinas, Campinas, Brazil (1998) (In portuguese)
38. Lima, A.G.B.: Diffusion phenomenon in prolate spheroidal solids. Case studies: Drying of banana. Doctorate thesis, State University of Campinas, Campinas, Brazil (1999) (In portuguese)
39. Lima, A.G.B., Queiroz, M.R., Nebra, S.A.: Simultaneous moisture transport and shrinkage during drying of solids with ellipsoidal configuration. Chem. Eng. J. **86**, 85–93 (2002)
40. Lima, A.G.B., Queiroz, M.R., Nebra, S.A.: Heat and mass transfer model including shrinkage applied to ellipsoidal products: case study for drying of bananas. Develop. Chem. Eng. Miner. Process. **10**, 281–304 (2002)
41. Lima, A.G.B., Farias Neto, S.R., Silva, W.P.: Heat and mass transfer in porous materials with complex geometry: Fundamentals and applications. In: Delgado, J.M.P.Q. (ed.) Heat and mass transfer in porous media. Springer, Berlin (2012)
42. Provenza, F.: Machine designer. Editora F. Provenza, São Paulo, p. 2.47, (1989) (In Portuguese)
43. Pólya, G., Szegö, G.: Inequalities for the capacity of a condenser. Am. J. Math. **LXVII,** 1–32 (1945)
44. Phoungchandang, S., Woods, J.L.: Moisture diffusion and desorption isotherms for banana. J. Food Sci. **65**, 651–657 (2000)
45. Dandamrongrak, R., Young, G., Mason, R.: Evaluation of pre-treatments for the dehydration of banana and selection of suitable drying models. J. Food Eng. **55**, 139–146 (2002)
46. Queiroz, M.R., Nebra, S.A.: Theoretical and experimental analysis of the drying kinetics of bananas. J. Food Eng. **47**, 127–132 (2001)
47. Demirel, D., Turhan, M.: Air-drying behavior of cavendish and gros Michel banana slices. J. Food Eng. **59**, 1–11 (2003)
48. Kaddumukasa, P., Kyamuhangire, W., Muyonga, J., Muranga, F.I.: The effect of drying methods on the quality of green banana flour. In African Crop Science Conference Proceedings, vol. 7, pp. 1267–1271. Kampala, Uganda (2005)
49. Talla, A., Puiggali, J.-R., Jomaa, W., Jannot, Y.: Shrinkage and density evolution during drying of tropical fruits: application to banana. J. Food Eng. **64**, 103–109 (2004)
50. Queiroz, M.R.: Theoretical and experimental study of the drying kinetics of banana. Doctorate thesis, State University of Campinas, Campinas, Brazil (1994) (In portuguese)

Solid State Fermentation: Fundamentals and Application

A. M. Santiago, L. S. Conrado, B. C. A. Mélo, C. A. B. Sousa, P. L. Oliveira and F. C. S. Lima

Abstract This chapter deals with the solid state fermentation theme, divided into two main parts: the first part presents information about the technology in solid state fermentation (SSF), emphasizing their definitions, advantages and disadvantages, the factors that influence this process, and the second part is about an experimental study using this technique in the production of pectinolytic enzymes. This experimental study evaluated the potential of dehydrated guava bark used as substrate in the production of the exo-polygalacturonase enzyme, obtained through solid state fermentation, using the microorganism *Aspergillus niger* as the fermenting agent. In this study, it was verified the influence of the initial moisture of

A. M. Santiago (✉)
Department of Chemical, Center of Science and Technology, State University of Paraiba (UEPB), Campina Grande, PB, Brazil
e-mail: angelasantiago@oi.com.br

L. S. Conrado · P. L. Oliveira
Academic Unit of Chemical Engineering, Federal University of Campina Grande, Campina Grande, PB, Brazil
e-mail: libiaconrado@yahoo.com.br

P. L. Oliveira
e-mail: palomalima_eq@hotmail.com

B. C. A. Mélo
Science and Technology Federal Institute of Education, IF Sertão-PE, Campus Petrolina, PE, Brazil
e-mail: beatriz.amorim@ifsertao-pe.edu.br

C. A. B. Sousa
Post-Graduate in Process Engineering, Center of Science and Technology, Federal University of Campina Grande, Campina Grande, PB, Brazil
e-mail: carlobispo@yahoo.com.br

F. C. S. Lima
Science and Technology, Federal Institute of Education – IFET/PE, Campus Belo Jardim, PE, Brazil
e-mail: flavia.c.7@hotmail.com

J. M. P. Q. Delgado and A. G. Barbosa de Lima (eds.), *Transport Phenomena and Drying of Solids and Particulate Materials*, Advanced Structured Materials 48,
DOI: 10.1007/978-3-319-04054-7_6,

the medium and the concentration of an additional source of nitrogen, at a process temperature of 30 °C, having as response the enzymatic activity. The experimental work showed that the maximum production of the enzyme was detected with initial moisture of the culture medium of 50 % (wb) and source concentration of nitrogen of 1.0 %, reaching a peak of activity of 12.64 U/g at 30 h of fermentation.

Keywords Pectinase · Guava peel · Fermentation

1 Introduction

The solid state fermentation (SSF) is a biotechnological process in which the growth of microorganisms will occur on and inside particles in solid matrix, and the liquid content contained in this matrix should be kept in values of water activity enough to ensure the growth and the cellular metabolism.

The literature reports a number of advantages arising from the application of this process, such as: the medium is easily aerated, provided there is space between the substrate particles (porosity), lower equipment cost, ease in extraction of the product, requires less space and, lower energy demand, can take advantage of agro-industrial waste (brans, peels and pomace) that are abundant low cost raw materials, and may be used as a carbon source in biotechnological processes for obtaining several products, including enzymes, alcohols, organic acids, amino acids, proteins, biologically active secondary metabolites and aroma compounds, provided that the chosen microorganism is suitable for the special purpose desired.

In Brazil, as agricultural activity is very intense and produces large amounts of organic residues, many research projects have been conducted using these residues for ethanol obtention, enzymes and protein enrichment since the 80s, adding value to tropical agricultural products and byproducts by solid state fermentation.

Many microorganisms can grow on solid substrates, agro-industrial residues; however, filamentous fungi are the most adaptable to the fermentation process in the solid state due to environmental conditions be closer to their natural habitat. These fungi are able to produce and excrete large amounts of substances, including enzymes, when degrading the carbohydrates found in vegetable biomass [1].

The solid state fermentation has gained importance and attention of researchers due to the low cost of the culture medium, high productivity and its recognition as a clean technology.

2 Solid State Fermentation

The occurrence of the practice of solid state fermentation comes from ancient times, when many people had already elaborated several forms of food using this microbial process, such as the use of soy sauce in China in 3000 BC, and in Japan and southwest Asia since 1,000 BC [2]. Although this process on solid medium has been developed for the manufacture of some foods and fermented beverages, currently its application has been extended to pharmaceutical and biochemical industries with emphasis on products such as enzymes, organic acids, antibiotics, surfactants, ethyl alcohol, biogas, microbial polysaccharides, biopesticides, and processes as protein enrichment of unconventional foods and pre-digestion of animal feed. However, the protein enrichment of animal feed was the main activity which motivated the use of SSF in Western countries after 1940, because this process allows the use of agro industrial residues, adding value to a low cost material, and in some cases minimizing the pollution caused by these residues [3].

The solid state fermentation (SSF), also called semisolid fermentation, can be defined as a technique for growth of microorganisms on and inside the moist porous particles (support or solid matrix), in which the content of liquid within the solid matrix should be kept in values of water activity that ensures the convenient growth and cellular metabolism, but not to exceed the maximum capacity of water retention in the matrix. The solid matrix can be, at the same time, support and substrate (organic and lignocellulosic materials or just a supporter which should be increased of nutrients) [4].

The commercial application of this process in Brazil is quite viable due to the availability of low cost agro industrial residues, resulting in a more environment friendly process, in addition to that, it generates products of interest to the food and pharmaceutical industries, and agriculture in general. Generally, it can be divided into two types: (a) socioeconomic applications such as waste composting, silage and utilization of agro industrial residues; (b) economically profitable applications such as the production of enzymes, organic acids and fermented foods [4].

A solid state fermentation (SSF) has the following features [5, 6]:

- The solid phase acts as a source of carbon, nitrogen and other components, besides serving as support for the growth of microbial cells;
- The air, required for the microbial development, must go through the empty spaces of the medium at relatively low pressures;
- Microbial growth occurs in conditions closer to the natural habitat;
- The medium shows high heterogeneity and the substrates are not completely accessible to the microorganism;
- The process is conducted without excess of free water, resulting in minimum water consumption, with low levels of residual water, also favoring the energy saving in the recovery process (downstream);
- The great possibility of using agricultural residues as a source of carbon and energy;
- The simplicity in the preparation of the fermentation medium since it usually requires only the main ingredient and water for moistening;
- Higher yields and high concentration of the desired product.

2.1 Advantages and Disadvantages of SSF

Among others, the main advantages of SSF are [7, 8]:

- The medium used is usually simple and consisted of unrefined agricultural wastes or products that can contain all the necessary nutrients for the development of the microorganism, which means that the use of a pre-treatment process is only a baking of the medium with water to moisten or dilate the substrate on the surface, which increases the accessibility of the microorganism to the internal nutrients, or simply a grinding the substrate into smaller particles;
- The disposal of the residues of this process is usually simple, because mostly the entire product is utilized, especially when the product is used as a supplement for animal feed;
- As the amount of water used in the process is small, the cost of the thermal process (sterilization) is reduced;
- For this reason, and because the substrate is concentrated, the space occupied by the fermentation equipment is small, considering the product yield;
- As most bacteria require high levels of liquid mixture, SSF excludes or reduces substantially the problem of bacterial contamination;
- Provided that there is space between the substrate particles, the solubility and diffusion of oxygen are greater;
- The remaining residue of this process has a low volume and is not able to develop pathogens;
- Besides the addition of water to moisten the medium, in some cases, it requires only the addition of nitrogen or mineral sources;
- It is possible to obtain high concentrations of spores at the end of the process, which is not possible in a submerged fermentation;
- Also according to these authors, the main disadvantages of the SSF are:
- As the SSF occurs under low free water, microorganisms that can be used in this process are limited. The most used ones are the fungi and some yeasts;
- In the case of operating the SSF on large scales, the heat generated by microbial metabolism must be removed, which is harder in this type of process than in submerged fermentation;
- When the size of the particles is too high, the oxygen transfer among the particles can become an issue;
- Monitoring of pH, O_2, CO_2 and the product yield calculation are more complex;
- Temperature control during fermentation is critical;
- Since in SSF the process takes place in a medium of high heterogeneity, control of microbial growth and variables such as agitation, aeration and concentration of nutrient and products is hampered.

2.2 Factors That Influence the Solid State Fermentation

There are several factors that contribute to the success of solid state fermentation process, as the amount of water available to the microorganism to develop and produce the desired bioproduct, temperature and pH, oxygen level, as well as the choice of raw materials to be used as substrate. This choice should be linked to the conditions related to the cost and availability of this material.

2.2.1 Moisture and Water Activity

The moisture of the medium determines the concentration of water contained in the material and is expressed in percentage terms. It is one of the factors that influences the most the fermentation process, because in cases that is too high, it will result in a decrease of the porosity of the substrate, limiting oxygen transfer inside the medium and consequently decreasing the gas exchange, besides increasing the risk of contamination. On the other hand, if it's less than needed by the microorganism, there will be greater difficulty in the nutrient diffusion, resulting in a lower growth of the microorganism and may influence the yield of the desired product [6].

The determination of its value in the process is closely related to the nature of the substrate, the needs of the microorganism used and the type of the final product desired [5, 6].

The moisture content in the solid state fermentation can vary between 18 and 85 %. This variation depends on the absorption power of the substrate, for example, the optimum initial moisture for the growth of the *Aspergillus niger* for the production of pectinase for the coffee pulp was 80 % [9], and for the passion fruit residue was 50 % [10].

The requirement of water by microorganisms is not associated with the solid substrate moisture, but to the water activity (a_w) of the medium, that is not function only of the water content, but also from substances which exert the bonding effect between the water and the food's structure. The relationship between the equilibrium moisture of a solid and the water activity at a given temperature is expressed graphically by means of sorption isotherms [11].

Water activity can be expressed as the ratio of the vapor pressure of the water contained in the substrate (P) and the vapor pressure of pure water (P_o), both at the same temperature according to the Eq. 1 [12].

$$a_w = \frac{P}{P_0} \quad (1)$$

where: a_w = water activity; P = vapor pressure of water contained in the substrate at a temperature T; P_0 = vapor pressure of pure water at the same temperature T of the substrate.

Every microorganism requires a minimum value, a maximum value and an optimal value of water activity to its growth and its metabolic activities. For filamentous fungi the minimum water activity is around 0.7 [13].

The enzymes require different minimum values of a_w on the substrate to perform its catalytic function. The minimum or the limit required for enzymatic activity varies from 0.25 to 0.70 for various oxidoreductases and 0.025–0.96 for several hydrolases [14].

A study designed to evaluate the influence of the initial moisture content of the medium (8.6–57.6 %) and water activity (0.641–0.990) in polygalacturonase by solid state fermentation of sunflower cake with different lineages of *Aspergillus* showed that the highest value of polygalacturonase activity found in wet basis was 38.36 U/g under the conditions of 51.8 % of the initial moisture and the water activity of 0.986 [15].

In the preparation of the medium in the fermentation processes it's necessary to consider its water activity and its moisture to achieve optimal levels of development of the microorganisms and consequently the maximum of the desired product. Therefore, for this it's necessary the knowledge of sorption isotherms, which can be defined as the graphic expression of the functional relationship between the content of the equilibrium moisture of a substance and water activity (a_w) of the same substance, measured at a constant temperature [16].

In the solid state fermentation the sorption isotherms can be constructed to evaluate the amount of water to be added to the substrate and also define the behavior of the residue with respect to addition of water.

Many mathematical models have been proposed for obtaining the adsorption/desorption isotherms of food, since there isn't a general equation, considering that the water activity depends on the composition and interaction of the different constituents of food with the water under conditions of thermodynamic equilibrium [17].

2.2.2 Temperature

Temperature has a great influence on the growth of microorganisms given that all fermentative processes are dependent on chemical reactions, which are affected by temperature. It is also considered a critical factor due to the accumulation of metabolic heat generated during the fermentation, which directly affects the germination of microorganisms and the product formation. In the composting process, this effect is desirable; however, to biotechnological processes such as the production of enzymes, the heat produced must be immediately dissipated, to avoid that the increase in temperature harms the desired fermentation [5, 6].

For each microorganism there is a minimum temperature below which there is no growth, an optimum temperature in which the growth is maximum and a maximum temperature above which there is no microbial development. For filamentous fungi the temperature directly influences spore germination, growth and formation of the products [6].

Many researchers studied the effect of different temperatures between 20 and 45 °C in pectinase production using *Aspergillus niger* as agent of the fermentation, and the temperature that favored the highest enzymatic activity was influenced by the substrate used, by the lineage of the microorganism used, but also depended on other operational factors such as type of reactor, the mass of the medium to be fermented, inoculum concentration, among others.

Since the increase in temperature influences negatively the kinetic of growth of the microorganisms, some studies suggest techniques for the removal of the heat generated during the fermentation process in the solid state on a large scale.

2.2.3 pH

The pH value of the medium is a very important parameter of any biological process in order to obtain a good cellular growth and high yields of products. An adverse pH affects at least two aspects of a living microbial cell: the operation of their enzymes and the transport of nutrients to inside the cell [18]. To minimize the effect of a sharp variation of pH, substrates with good buffering capacity are generally used or buffer solutions are added during the step of moistening of the material [5].

Microorganisms have minimum, optimum and maximum pH values for their development and growth. Generally the fungi prefer pH between 3.5 and 5.0 and the bacteria pH close to neutrality (6.5–7.0). To the Aspergillus niger the minimum value is 1.2 and the optimum is between 3.0 to 6.0 [19].

The study of thermal stability and stability against variations of the pH of polygalacturonase produced by *Thermoascus aurantiacus* showed that between pH 3.0 and 3.5 the enzyme was maintained at 91 % of its original activity, and in high pH values there was a decrease in stability [20]. With respect to temperature, for values up to 50 °C, this enzyme retained 80 % of its original activity and only maintained 6 % of this activity when exposed to a temperature of 70 °C. However, other work has also studied the thermal stability and stability against variations of the pH of polygalacturonase produced by solid state fermentation, using the microorganism *Aspergillus niger* and as substrate the residue of the passion fruit showed that this enzyme presented good stability up to 50 °C, presenting 75 % of maximum activity at 30 °C. At temperatures above 70 °C it was completely inactivated, during 20 min and with respect to pH the authors observed that this enzyme showed good stability at pH values between 3.5 and 5.5 and above 6.5, did not detect activity [21].

A study to evaluate the thermal stability of polygalacturonase enzyme produced in cashew bagasse showed it remained at 90 % of its activity at 40 °C and the optimum pH was between 2.5 and 3.5 [22]. For most industrial processes, the polygalacturonase produced by fungi are used in the low pH range.

2.2.4 Substrates Used in Solid State Fermentation

Knowledge of the physicochemical characteristics of solid substrates to be used as a culture medium for cellular growth and product yield is of fundamental importance for their choice, in addition to the conditions of cost and availability.

The substrates used should contain sources of carbon and energy, nitrogen sources, minerals, and water. However, for the production of inducible enzymes, the presence of an inducer substance is essential, for example, pectin for pectinase [23].

The choice of the substrate is as important in the cultivation in solid state as the choice of the microorganism. Not always the medium that allows the best development of microorganism favors a higher amount of enzymes. The appropriate balance between nitrogen sources is as important to the nutritional requirement of the microorganism as the effects of environmental conditions that affect the mycelial growth [24].

The use of agricultural residues or byproducts from food industries as substrate in solid state fermentation is increasingly widespread, seeking to obtain products of high commercial value and low cost of production, besides minimizing its accumulation in the environment. These residues are the most researched substrates in this process, due to its abundance, cost, besides being rich in nutrients that promote the growth of microorganisms and formation of the desired product [16].

There are several agro-industrial waste used as substrate in solid state fermentation, among these are: orange bagasse [25], lemon peel [26], pineapple bagasse [27], wheat bran and rice husks [28], coffee pulp [9], peel and albedo of passion fruit [29], bran wheat [30], passion fruit peel [31], grape pomace [32], lemon peel [33]. The research groups that study the production of pectinase enzymes generally take advantage of regional substrates.

In the case of citrus fruits, the peel is the main byproduct generated in its industrial processing, constituted approximately about 50 % of weight of fresh fruit [26]. The destiny of this byproduct represents a challenge for many factories, however the microorganisms easily utilize this substrate in fermentations due to its rich composition (80 % organic material, 19–30 % of free sugars, 20 % pectin and 57 % of dietary fiber) [34]. Pectin is a suitable substrate for the growth of fungi and production of pectinolytic enzymes by microbial systems.

For a better utilization of substrates by microorganisms some kind of pre-treatment is necessary, considering that they are not capable of delivering macromolecular structures through the cell membrane. This pre-treatment can be: asepsis of the material for reducing microbial load, crushing or grinding to reduce the particle size; nutrient supplementation (nitrogen, phosphorus, etc.) to the solid substrate by adapting it better to the nutritional conditions of the microorganism; adjustment of pH and moisture content, and a steam treatment to pre-degradation of macromolecular structure and elimination of the main contaminants [35].

The addition of water or nutrient solution to the culture medium should be made so to achieve ideal levels of water activity for the development of microorganism under study.

2.2.5 Microorganism

The proper selection of the microorganism is one of the most important criteria to consider in a semisolid culture. Among the microorganisms, the filamentous fungi, the main producers of commercial enzymes, are the most important and best adapted to this process. Its growth form in hyphae, and good tolerance to low water activity and high osmotic pressure confers advantages over other microorganisms, bacteria and yeasts, to the fungi; in addition to the conditions of the culture medium are closer to their natural habitat [36].

The filamentous fungi are easy to handle and grow at low pH, the risks of bacterial contamination in the environment in which they are located are very small, they easily sporulate and have the ability to produce and excrete large quantities and varieties of specific products, including pectinase enzymes. In general, these fungi play an important role in food industries because they have different types of enzymes capable of metabolize complex mixtures of organic compounds present in most residues [37].

In solid state fermentation the fungal growth begins on the substrate surface through its mycelial structure, which has hyphae that are able to penetrate the pores of the substrate, benefiting in its interior the specific metabolic reactions and consequently favoring the colonization and utilization of nutrients available in the medium, thus generating its bioproducts.

The *Aspergillus niger*, a black filamentous fungus generally called "black mold", features thin hyphae, septate and conidiophores with vesicles covered with black conidium that can be visualized by electron micrograph. This fungus is quite commonly used in fermentation processes due to its ease of handling, ability to ferment a wide variety of low cost raw materials and produce high yields of bioproducts [18]. Mutant forms of this species are also quite commonly used in fermentation processes in order to synthesize higher amounts of enzymes.

The culture of *Aspergillus niger* has the ability to produce a variety of enzymes, depending on the induction and/or the substrate used.

2.2.6 Inoculum

It is important to quantify the concentration of inoculum in the fermentation process, because a low concentration can promote the growth of undesirable contaminants and form little biomass, although an inoculum with a high concentration of spores will be able to exhaust the medium for biomass formation, decreasing the amount of product desired.

A study of the influence of inoculum concentration of the *Aspergillus niger* (CCT 0916) using the dry bagasse from the cashew stalk as substrate in solid state fermentation for the production of polygalacturonase obtained the biggest polygalacturonase activity of 33 U/g with spore concentration of 10^6 spores/g of wet medium [38].

Another study in this direction, evaluated the influence of inoculum concentration of *Aspergillus niger* (3T5BS) and moisture of the semisolid culture medium on the production of polygalacturonase in aerated columns, varying the relation substrate/solution in 0.06 and 0.08 L for each 100 g of medium and the inoculum concentration of 10^5 and 10^7 conidia/g bran. The authors concluded that the relation of 0.06 and the conidia concentration of 10^7 were the most favorable conditions for the synthesis of polygalacturonase with maximum production of 33.08 U/mL at 64 h of fermentation [39].

2.2.7 Particle Size

The size and shape of the particles of the substrate influence the solid state fermentation, since usually very small particles may result in its agglomeration interfering with microbial respiration and consequently little cellular growth, however the larger particles promote better aeration due to spaces between the particles, but they limit the surface to microbial attack [40].

It is also worth noting that in very small particles, the surface area is larger while the porosity is smaller, hindering the penetration of the filamentous fungus in the pores and within of the substrate particles [41].

2.2.8 Agitation

Agitation facilitates maintaining the biomass more homogeneous as possible, as the distribution of water of the medium, nutrients, inoculum, as well as the temperature and aeration in the bioreactor during the semisolid culture. The care in this step is with the fractionation of mycelium and may interfere in the formation of spores and in the development of microorganism, if it occurs in uncontrolled speed.

3 Production of the Enzyme Exo-polygalacturonase by SSF

There are countless applications of solid state fermentation, the most important being the production of enzymes, organic acids, antibiotics, surfactants, ethyl alcohol, biogas, microbial polysaccharides and biopesticides.

Enzyme production is a biotechnological area of great commercial interest, given that sales of this product in the whole world revolve around billions of dollars annually and the solid state fermentation is considered a great system for producing them in Brazil, considering the huge generation of agro-industrial residues.

Among the enzymes produced by this technique, pectinase stand out because they represent about 25 % of the world market for enzymes, being widely used in fruit juice industries to reduce viscosity and consequently increasing the efficiency of the extraction and clarification, the maceration of the grapes into wine, the extraction of vegetable oils increasing its yield and reducing its processing time and treatment of natural fibers in the textile and paper industry [42, 43].

Pectinases are a heterogeneous group of enzymes that hydrolyze the pectic substances and are able to recognize the glycosidic bonds of the type alpha-1.4 between units of galacturonic acid or its derivative methoxylated. They are produced by plants and microorganisms and its substrate are the constituent polysaccharides of the middle lamella and primary wall of plant cells. However, the fungi are the favorite in industrial scale, because approximately 90 % of the enzymes produced by them can be secreted into the medium, besides producing large quantities [44, 45].

These enzymes can be produced both by submerged fermentation (SF) and by solid state fermentation (SSF), however, the SSF is generally considered the most susceptible technique to higher yields of pectinases [46].

The research groups that study the production of pectinase enzymes generally tend to take advantage of regional substrates. In this study, we used guava residue since Brazil is the biggest producer of red guavas in the world. Most of this production is destined to the industry for the production of juice, frozen pulp, sweets, jam, ice cream and nectar, generating about 30 % of residues, which has the potential to be used in bioprocesses. Thus, it will be explained a study about the production of pectinolytic enzymes using the guava residues as raw material, and with *Aspergillum niger* as the metabolizer microorganism. This work aimed to study the production of pectinase by solid state fermentation with flour from guava peel, evaluating the influence of moisture of the medium and the concentration of the additional nitrogen source on the enzymatic activity, expressed in exo-polygalacturonases.

The trials of this work were performed at the Laboratory of Biochemical Engineering (LEB) from the Academic Unit of Chemical Engineering (UAEQ) of the Federal University of Campina Grande/PB (UFCG).

3.1 Raw Material

The raw material used as substrate in solid state fermentation was the guava peel (*Psidium guajava*) of the Paluma variety acquired in the EMPASA (Empresa Paraibana de Abastecimento e Serviços Agrícolas) in the city of Campina Grande-PB according to the maturation stage, popularly known as "de vez" in order to obtain more homogeneous samples and substantial amounts of pectin.

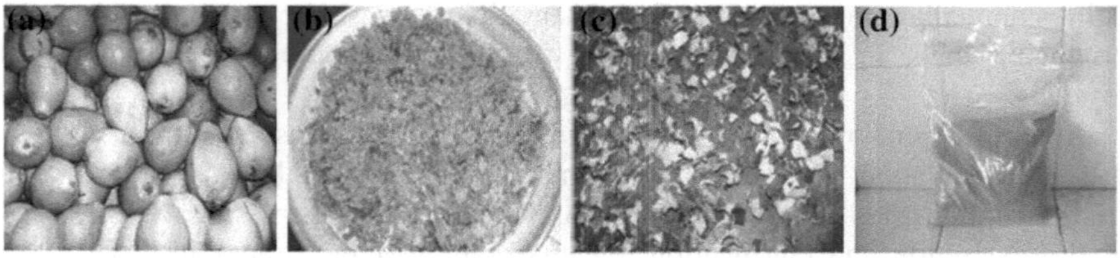

Fig. 1 Stages of preparation of the flour from guava peel (dry residue): whole fruit (**a**), fresh peels (**b**), dry peels (**c**) and the milled peels (**d**)

3.2 Preparation and Characterization of the Flour from Guava Peel

The fruits were transported to the laboratory, and they were washed in chlorinated water at 2.5 % and rinsed with running water. Then they were subjected to the process of selection, sorting and peeling. The peels were placed in aluminum trays in a greenhouse with forced air circulation at the temperature of 55 ± 1 °C to constant mass. After drying they were ground in a cutting mill to obtain the appropriate granulometry to the solid state fermentation process for the production of the enzyme. Then they were homogenized and submitted to the quartering technique for the removal of approximately 150 g to perform the physicochemical characterization.

The determinations of pH, moisture, ash and soluble solids (SS) followed the standards Brazil [47]. The amount of pectin was determined by gravimetric precipitation method using calcium pectate, described by Rangana [48]. The reducing sugars (RS) and total reducing sugars (TRS) were determined using 0.5 g of the sample by DNS methodology [49] in spectrophotometer with glucose solution as standard. The crude protein content was determined by semi-micro Kjeldahl method with adaptation to nitrogen (N). by UV-visible spectrophotometry at 410 nm, following the method described in Le Poidevin and Robinson [50]. The water activity was determined using the equipment Novasine Thermoconstanter (RTD-200 TH2) at temperature of 30 °C. The analyses were performed in triplicate. The remaining of the flour from guava peel was stored in a hermetically sealed glass container at room temperature for later use as raw material in the fermentation process.

Figure 1 shows the whole fruits (a) the fresh peels (b) the dry peels (c) and the milled peels (d) which were used in this work.

3.2.1 Moisture Adsorption Isotherms of the Flour from Guava Peel

For the construction of the isotherms it was weighed 20 g of the dry residue and it was added distilled water in the proportion of 1:1. Then it was submitted to

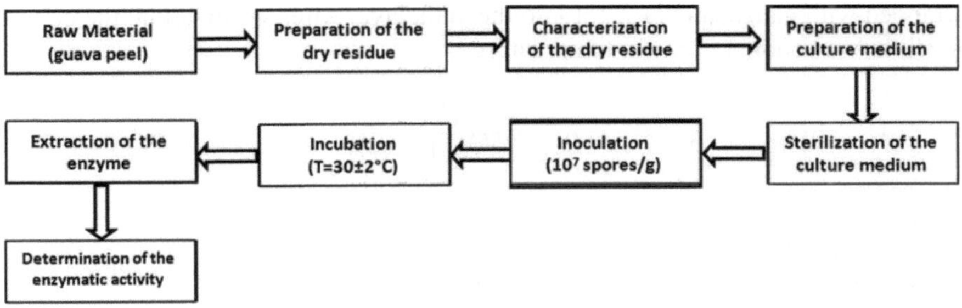

Fig. 2 Stages of enzyme production by solid state fermentation

homogenization for 30 min and allowed to stand in a sealed container in the refrigerator for 24 h. Completed the period of rest of the dry residue, was collected 1 g of moist material, and transferred to aluminum capsules, which were to the equipment Thermoconstanter Novasina (RTD-200 TH2) to perform the readings of water activity at temperatures of 25, 30 and 35 °C. After the reading in the equipment, the capsules containing the residue were placed in a greenhouse at 105 °C, they were cooled in a desiccator and again brought to Novasina for reading the water activity. This experiment was conducted to achieve the mass balance of the sample, and finally the same were brought to the greenhouse to determine the dry mass. The moisture adsorption isotherms were constructed with the data of equilibrium moisture and the water activity.

3.3 Stages of the Fermentation Process

The solid state fermentation was performed in 250 ml Erlenmeyer flasks in which were put quantities of the moistened residue with distilled water reaching moisture 40, 50 and 60 % and the concentration of the nitrogen source (ammonium sulfate) of 0.5, 1.0 and 1.5 %. Figure 2 illustrates the stages of the solid state fermentation process.

3.3.1 Preparation of the Culture Medium

The culture medium used was flour from guava peels moistened with distilled water according to a_w (water activity) required for the metabolism of the micro-organism. The amount of water to be added to the medium was calculated according to Eq. 1. The nitrogen source used was ammonium sulfate, which was diluted in the amount of distilled water to be added to the medium and after the homogenization and the moistening of the medium, was distributed 10 g in each Erlenmeyer flask of 250 ml capacity, which were stoppered with cotton plug wrapped with gauze, and brought into the autoclave. After sterilization, the culture medium was cooled to room temperature.

$$m_{H_2O} = m_{RES}\frac{(U_2 - U_1)}{(1 - U_2)} \quad (2)$$

where:

m_{H_2O} mass of water required to hydrate the medium;
m_{RES} mass of the dry residue used in the fermentation;
U_1 moisture present in the dry residue;
U_2 moisture required by the fermentation process.

3.3.2 Microorganism

The microorganism used was *Aspergillus niger* mutant modified CCT 0916, donated by Empresa Brazileira de Pesquisa Agropecuária (EMBRAPA, Fortaleza-Brazil) and maintained in the Laboratory of Biochemical Engineering (LEB/UAEQ/UFCG) in test tubes with screw caps containing sterile soil, maintained at −18 °C.

3.3.3 Inoculation and Incubation

Initially, the microorganism was cultured on corn for obtaining a large amount of spores. Then the spores were dispersed with the assistance of 40 ml of thickening solution of Tween 80 at 0.3 % v/v. The concentration of spores in the suspension obtained was quantified by counting under a microscope using a mirrored Neubauer Chamber after the appropriate dilution of the suspension obtained. The volume of the spore suspension added to the fermentation medium was adjusted so as to obtain a concentration of 10^7 spores per gram of solid substrate. After the inoculation of spores in the sterilized medium, the bioreactors (Erlenmeyer flasks) were incubated in a greenhouse at (30 ± 1) °C for 72 h while maintaining the moisture of the incubation environment to a kinetic monitoring. Seven samples were prepared for each assay in order to remove them periodically to determine the following analyzes: moisture, pH, reducing sugars and the exo-polygalacturonase activity (PG act).

3.3.4 Extraction of the Enzyme and Determination of Enzyme Activity

The extraction of the enzyme complex was performed by adding to the Erlenmeyer flasks containing the fermented medium the quantity of 5.0 ml of acetate buffer

Table 1 Physicochemical characterization of the flour from guava peel (dry residue)

Parameters Analyzed	Dry residue
Moisture (wb) (%)	12.36 ± 0.51
Ash (%)	3.89 ± 0.02
Protein (%)	2.09 ± 0.20
Pectin (%)	9.92 ± 0.13
Total reducing sugars (%)	26.51 ± 1.06
Reducing sugars (%)	23.35 ± 0.19
Total soluble solids (Brix)	30.00 ± 0.00
pH	4.24 ± 0.01
Apparent specific mass (ρ_a) (g/mL)	0.55 ± 0.04
Real specific mass (ρ_r) (g/mL)	1.20 ± 0.05
Porosity of the bed of particles	0.54

200 mM pH 4.5 for each gram of the fermented medium (Extraction ratio (ER) = 5 ml/g) which were left to stand for 1 h in the thermostatic bath at 30 °C. The crude enzyme extract obtained was filtered through filter paper (Whatman 1) and then was determined the activity of the polygalacturonase enzyme (*PG act*) according to Couri and Farias [51], following the adaptation of EMBRAPA, Fortaleza-Brazil. One unit of polygalacturonase activity was defined as the amount of enzyme that liberates 1 μmol of galacturonic acid per minute on the reaction conditions.

3.4 Results and Discussion

3.4.1 Physicochemical Characterization of the Flour from Guava Peel (Dry Residue)

In a fermentation process the starting point is the characterization of the material that will be used as culture medium to obtain the desired product.

The data in Table 1 correspond to physicochemical characterization of the dry residue on a wet basis.

The dry residue studied showed a high content of pectin (9.92 %), which according to the literature this chemical substance acts as an efficient inducing source in the production of pectinolytic enzymes as well as also being able to act as a carbon source for microorganisms, depending on amount of sugars present in the culture medium. The pectin content is influenced by several factors, among them the maturity stage of the fruits.

Fontana et al. [52] studied the effect of the amount of pectin in a medium based with wheat bran and observed that up to 16 % there were no repression of the synthesis of pectinases produced by *Aspergillus niger* with this amount of this substance.

For the reducing sugar the values found were 23.35 %, while Munhoz et al. [53] found 9.57 %. However, the values of total reducing sugars were 26.51 % close to that found by Munhoz et al. [53] which was 30.64 %. The amounts of reducing sugars (RS) and of total reducing sugars (TRS) found differ from those of the cited authors due to several factors, including the maturation state of the fruit, because in this study was used only fruit "de vez" due to the pectin content being larger, as well as also the washing carried out with this residue during the stage of preparation.

The pH found is close to the value cited in the literature. This parameter has great influence on solid state fermentation, given that the growth of microorganisms depend on the initial pH value of the medium, which for the *Aspergillus niger* the minimum is 1.2 and the optimum is between 3.0 and 6.0.

Malvessi and Silveira [54] used a medium containing wheat bran, mineral salts and pectin in the production of the polygalacturonase enzyme varying the pH from 2 to 7 and obtaining a maximum activity with initial pH of the medium of 4.0.

The apparent specific mass 0.55 g/ml reveals that has no tendency to compact itself, creating empty spaces between the particles of the same and consequently favoring the respiration and metabolism of the microorganisms.

Adsorption Isotherms of the Flour from Guava Peel (Dry Residue)

The moisture adsorption isotherms of the dry residue from guava were built at temperatures of 25, 30 and 35 °C. Through these it was possible to verify the content of free water contained in the dry residue as well as determining the range of moisture that should be fermented to correspond to the favorable water activity to the development of the fungus *Aspergillus niger* CCT 0916.

Modeling of the Moisture Adsorption Isotherms

Figure 3 shows the adsorption isotherms moisture at temperatures of 25, 30 and 35 °C of the dry residue from guava with the respective values of equilibrium moisture (U_{eq}) in function of water activity (a_w).

It is observed that the equilibrium moisture (U_{eq}) increases with the increase of water activity (a_w). According to LEITÃO [13], the microorganism *Aspergillus niger* develops in culture mediums with minimal water activity of 0.70. This way, the guava residue should have its equilibrium moisture adjusted to at least 0.38.

There is in the literature a variety of mathematical equations that describe the sorption isotherms among them the Guggenheim-Anderson de Boer (GAB) that are fairly applied in food, so through the Software Statistica 5.0, were performed adjustments of adsorption isotherms using the GAB model.

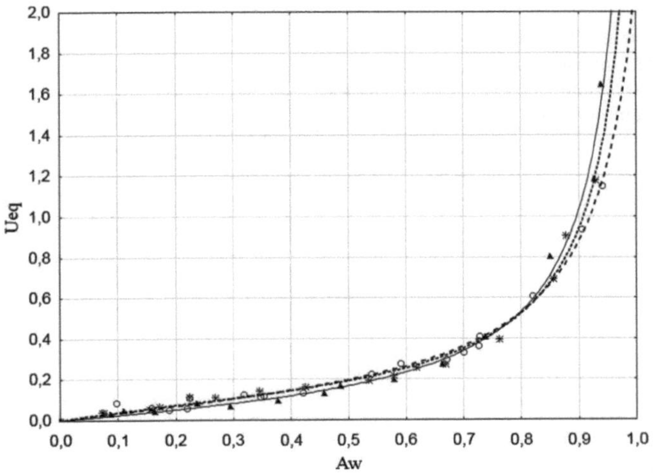

Fig. 3 Adjust of the adsorption isotherms of guava flour (*black circle* 25 °C, *black triangle* 30 °C and *asterisk* 35 °C)

3.4.2 Solid State Fermentation Process

In Figs. 4, 5, 6 and 7, we observe the kinetic monitoring of exo-polygalacturonase production and consumption of reducing sugars (RS) as well as monitoring of pH and moisture on a wet basis (wb) during the period of 72 h in solid state fermentation for each studied condition. The concentrations of nitrogen used in the medium varied from 0.5 to 1.5 % and the initial moisture of 40 and 60 % (wb) corresponding to water activity (a_w) of 0.77 and 0.83, respectively.

Comparing the trials it is noticed that the ones with lower initial moisture of the medium are the ones with lower polygalacturonase activity (PG act), probably the low moisture caused modifications in the cellular membrane of the fungus conducing the transport limitations and affecting the microbial metabolism. As the initial moisture of the fermentation medium is a factor synthesis inhibition of PG, the important is to work with the moisture content that maintain the balance between the availability of water, expansion of the substrate and effects of oxygen diffusion.

The consumption of reducing sugars was faster in trial 4 coinciding with the maximum value of activity. This fact shows that the product formed is usually directly connected to the decomposition reactions of the substrate (sugars).

To check the data reproducibility were performed trials in triplicate under the following conditions: initial moisture of the medium of 50 % (wb), corresponding to the water activity (a_w) of 0.80 and concentration of the nitrogen source at 1.0 %, according to Fig. 8.

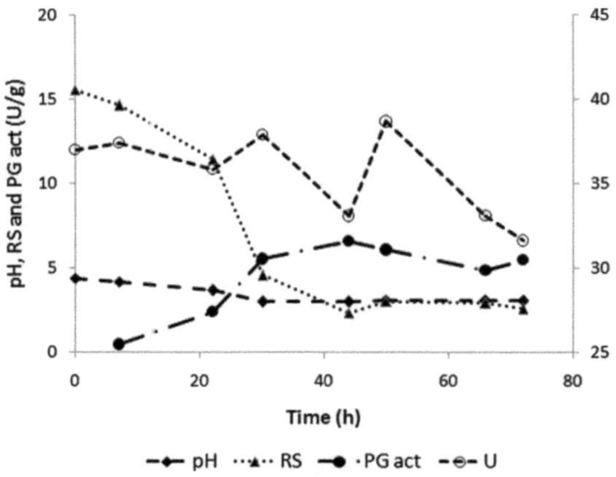

Fig. 4 Solid state fermentation of the flour from guava peel with moisture of the medium 40 % and nitrogen concentration of 0.5 %

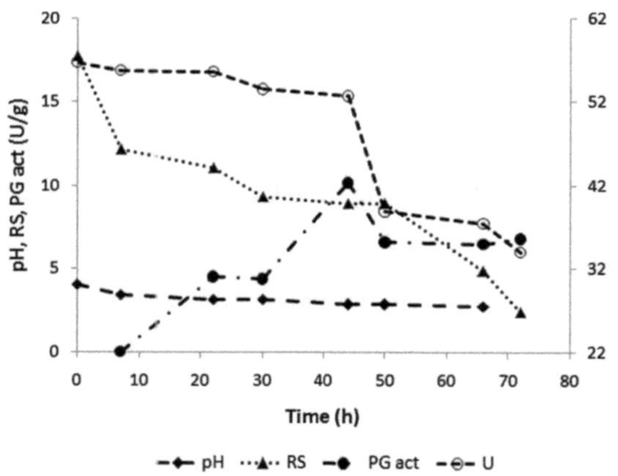

Fig. 5 Solid state fermentation of the flour from guava peel with moisture of the medium 60 % and nitrogen concentration of 0.5 %

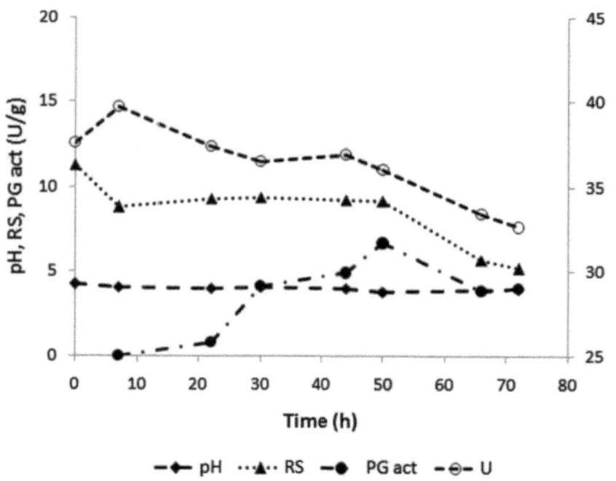

Fig. 6 Solid state fermentation of the flour from guava peel with moisture of the medium 40 % and nitrogen concentration of 1.5 %

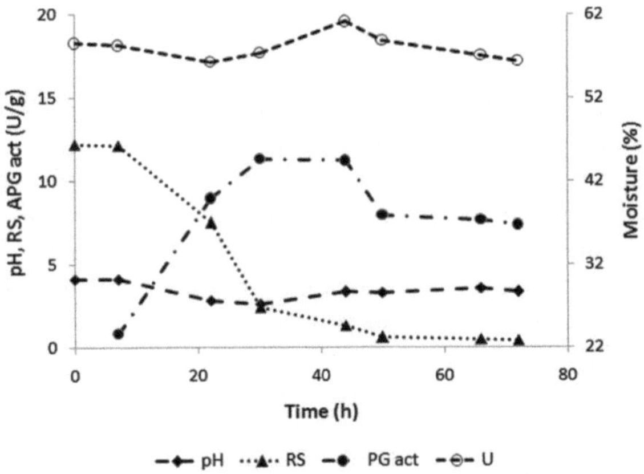

Fig. 7 Solid state fermentation with moisture of the medium 60 % and nitrogen concentration of 1.5 %

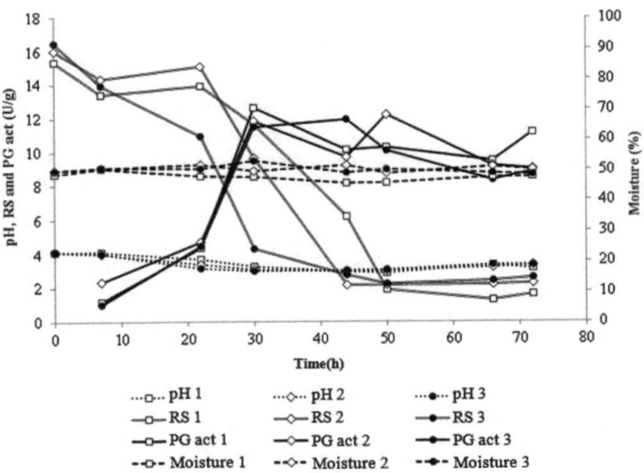

Fig. 8 Solid state fermentation of the flour from guava peel with moisture of the medium 50 % and nitrogen concentration of 1.0 % (analysis of the reproducibility of the experimental data)

In three trials the consumption of reducing sugars (RS) by microorganisms were similar, starting with the content of approximately 16 %, reaching the lowest value of 2 % at 66 h of fermentation. The maximum polygalacturonase activity was 12.64 U/g at 30 h of culture.

4 Conclusions

The theoretical study of solid state fermentation showed that this is a viable and promising technique in the development of bioproducts and bioprocesses, among them the production of enzymes, provided that the conditions of the culture medium and microorganism used are appropriate to raw materials to be fermented.

The experimental study it was noticed that the flour guava bark is consisted of easily metabolizable chemical substances by microorganisms such as reducing sugars, besides containing a high amount of pectin, which is an inducing source in the production of pectinolytic enzymes and the maximum activity of the exo-polygalacturonase enzyme was given with initial moisture of the culture medium of 50 % (wb) and the concentration of nitrogen source of 1.00 %, reaching a peak of activity of 12.64 U/g of the medium fermented at 30 h of fermentation.

References

1. Castilho, L.R., Medronho, R.A., Alves, T.L.M.: Production and extraction of pectinases obtained by solid state fermentation of agro-industrial residues with *Aspergillus niger*. Bioresour. Technol. **71**, 45–50 (2000)
2. Araújo, L.F.: Enriquecimento Protein Mandacaru without spines and Palma Forager by semi-solid fermentation. 2004. Ph.D. thesis in process engineering—Post-Graduate Program in process engineering. Federal University of Campina Grande, Paraíba (2004)
3. Pandey, A.: Solid-state fermentation. Biochem. Eng. J. **13**(2), 81–84 (2003)
4. Palma, M.B.: Production of xylanases by Thermoascusaurantiacus in solid state cultivation. 2003. 169f. Ph.D. thesis in chemical engineering. Federal University of Santa Catarina (2003)
5. Del Bianchi, V.L., Moraes, I.O., Capalbo, D.M.F.: Solid state fermentation. In: Schmidell, W., Lima, U.A., Aquarone, E., Borzani, W. (eds.) Industrial Biotechnology: Biochemical Engineering, vol. 2. Edgard Blucher Ltda., São Paulo (2001)
6. Pinto, G.A.S., Brito, E.S., Silva, F.L.H., Santos, S.F.M., Macedo, G.R.: Solid state fermentation; one alternative to the use and enhancement of agro-industrial waste. J. Ind. Chem. **74**, 17–20 (2006)
7. Bramorski, A.: Characterization of growth and the production of volatile metabolites of filamentous fungi cultivated on substrates agribusiness. Master thesis, Federal University of Paraná (1997)
8. Lu, M.Y., Maddox, I.S., Brooks, J.D.: Aplication of a multi-layer packed-bed reactor to citric acid production in solid state fermentation system: a review. Process Biochem. **33**(2), 117–123 (1998)
9. Orozco, A.L., Pérez, M.I., Guevara, O., Rodríguez, J., Hernández, M., González-vila, F.J., Polvillo, A.M.E.: Biotechnological enhancemente of coffe pulp residues by solid-state fermentation wuthstreptomyces Py-GC/MS analysis. J. Anal. Appl. Pyrol. **81**, 247–252 (2008)
10. Alcântara, S.R., Silva, F.L.H.: Moisture, pH, reducing sugars and poligalacturonásica activity in a process of solid state fermentation using tray reactor. In: 14th Brazilian Congress of EngenhariaQuímica (COBEQ), Búzios (2012)
11. Labuza, T.P.: Sorption phenomena in foods. Food Technol. **22**(3), 263–272 (1988). Quoted by: Prado, M.E.T. et al.: Sorption isotherms of dates: experimental determination and evaluation of mathematical models. Food Sci. Techonol. **19**(1). Campinas (1999)

12. Fellows, P.J.: Teconologia of Food Processing: Principles and Pratice. Fennema, 2nd edn, p. 602. Artmed, Porto Alegre (2006)
13. Leitão, M.F.F.: Water activity and microbiological changes of food. In: Garden, D.C.P, Germer, S.P.M. (eds.) Water Activity in Foods, pp. 1–8. ITA, Campinas (1997)
14. Fennema, O.R., Damodaran, S., Parkin, K.: Chemical Alimentos, 4th edn, p. 900. Artmed, Porto Alegre, (2010)
15. Castro, R.J.S., Freitas, A.C., Beserra, M.A., Vieira, J.M.M., Pinto, G.A.S.: Production of polygalacturonase from Aspergillus niger in solid state fermentation using as substrate rapeseed cake. In: em fermentação semi-sólida utilizando como substrate torta de canola. In: National Symposium Bioprocesses, Anais…Natal/RN, Natal/RN, 2–5 Aug 2009
16. Correia, R.T.P. Study of cultivation in semisolid residue pineapple by Saccharomyces cereviseae e Rhizopus oligosporus. 138f. Ph.D. thesis in chemical engineering, Post-Graduate Program in Chemical Engineering, Federal University of Rio Grande do Norte, Natal (2004)
17. Welti-Chanes, J, Vergura, B.F.: Actividad de água. Concepto y aplicacióninn foods with high content of humedad. In: Aguilera, J.M. (ed.) Issues in Technology Alimentos, vol. 1, pp. 11–43. México (1997)
18. Jay, J.M.: Microbiology of Food, 6th edn, p. 709. Artmed, Porto Alegre (2005)
19. Franco, B.D.G.M.: Food Microbiology, p. 182. Publisher-Atheneu, São Paulo (1999)
20. Martins, E.S.: Purification of thermostable polygalacturonases produced by the fungus Thermoascusaurantiacus through submerged fermentation and solid state fermentation and biochemical characterization of the same. 132f. Ph. D. thesis in biosciences, Universidade Estadual Paulista Júlio de Mesquita Filho, Rio Claro, São Paulo, (2006)
21. Souza, R.L.A., Conrado, L.S.O., Silva, F.L.H., Amorim, B.C.: Characterzation of polygalacturonase produced by solid state fermentation using the passion fruit residue as substrate. Braz. J. Agro-industrial Prod. **14**(9), 987–992 (2010)
22. Leite, N.J., Alcântara, S.R., Lima, F.C.S., Silva, F.L.H.: Study of the stability of the enzyme polygalacturonase against pH and temperature. In: Symposium on Science and Food Technology, Anais, Joao Pessoa, 12–14 Nov 2012
23. Bon, E.P.S., Ferreira, M.A.: Enzymes in Biotechnology Production, Application and Market, p. 506. Interscience, Rio de Janeiro (2008)
24. Raimbault, M., Deschamps, A., Meyer, F., Senez, J.C.: Direct protein enrichment of starchy products by fungal solid fermentation. In: Proceedings of Giam-V, Marseilles (1977)
25. Camargo, L.A., Dentillo, D.B., Cardeiro, L., Gattás, E.A.L.: Use of Orange bagasse in the production of pectinase from Aspergillus sp. Alim. Nutr. Araraquara **16**(2), 153–156 (2005)
26. Patil, S.R., Dayanand, A.: Production of pectinases from deseeded sunflower head by Aspergillus nigerin submerged and solid-state conditions. Biores. Techonol. **97**, 2054–2058 (2006)
27. Imandi, Z.S.B., Bandaru, V.V.R., Somalanka, S.R., Bandaru, S.R., Garapati, H.R.: Aplication of statistical experimental designs for the optimization of médiumconstituintes for the production of citric acid from pineapple waste. Bioresour Technol. **99**, 4445–4450 (2008)
28. Nizamuddin, S., Sridevia, A., Narasimha G.: Production of β-galactosidase by *Aspergillus oryzae* in solid-state fermentation. Afr. J. Biotechnol. **7**(8), 1096–1100 (2008)
29. Souza, R.L.A.: Pectinase production by solid state fermentation using as substrate residue of passion. 97f. Dissertation (Masters in Chemical Engineering), Post-Graduate Program in Chemical Engineering, Federal University of de Campina Grande, Campina Grande, PB (2008)
30. Okafor, U.A., Okochi, V.I., Shalom, N.C., Ebuehi. O.A. T., Okerenta, B.M.: Pectinolytic activity of wild-type filamentous fungi fermented on agro-wastes. Afr. J. Microbiol. **4**(24), 2729–2734 (2010)
31. Rocha, C.P., Bailão, E.F.L., Coutinho, F.U., Cardoso, V.L.; Use of agro-industrial residues as substrate for the production of amylase and pectinase by Aspergillus niger. In: 18th National Symposium Bioprocesses, pp. 24–27. Proceedings….Caxias do Sul. Anais…Caxias do Sul/ RS, CD-Rom (2011)

32. Díaz, A.B., Ory, I., Ildefonso, C., Blandini, A.: Enhance hydrolytic enzymes production by *Aspergillus awamori*on suplemented grape pomace. Food and Bioprod. Process. **90**, 72–78 (2012)
33. Ruiz, H.A.; Rodríguez, J., Rodríguez, R., Contreras, J.C., Aguilar, C.N., Pectinase production from lemon peel pomace as support and carbon source in solid-state fermentation column-tray bioreator. Biochem. Eng. J. **65**, 90–95 (2012)
34. Chau, C.F., Huang, Y.L.: Comparison of the chemical composition and physicochemical properties of different fibers prepared from the peel of *Citrus sinensis I. Cv. Liucheng*. J. Agric. Food Chem. **51**, 2615–2618 (2003)
35. Mitchell, D.A., Berovic, M., Krieger, N.: Biochemical engineering aspects of solid state bioprocessing. Adv. Biochem. Eng. Biotechnol. **68**, 61–138 (2000)
36. Rocha, C.P.; Otimization of enzyme production by Aspergillus niger in solid state fermentation. 98 f. Dissertation (Masters in Chemical Engineering), Post-Graduate Program in Chemical Engineering, Federal University of Minas Gerais-Uberlândia, MG (2010)
37. Pelczar, J., Chan, E.C.S., Michael, J., Krieg, N.R.: Microbiology: Concepts and Applications, vol. 1, 2nd edn, p. 523. Mmakron Books, São Paulo (1996)
38. Alcântara, S.R, Silva, F.L.H.: Influence of spores concentration, moisture, ammonium sulphate concentration and temperature on polygalacturonase production using cashew apple in the solid state fermentation process. Chem. Eng. Trans. **24**, 949–954 (2011)
39. Terzi, S.C., Carvalho, C.V.P., Oliveira, A.C.P., Couri, S.: Influence of varying the concentration of inoculum of *Aspergillus niger* 3T5B8 and humidity of fermentation médium on enzyme production polygalacturonase. In: 10th National Symposium Bioprocesses, Anais…Frorianópolis/SC, Florianópolis (2003)
40. Pandey, A., Selvakumar, P., Soccol., Nigam, P.: Solid state fermentation for the production of industrial enzymes. Curr. Sci. **77**, 149–162 (1999)
41. Ruiz, H.A., Rodríguez, J., Rodríguez, R., Contreras, J.C., Aguilar, C.N.: Pectinase production from lemon peel pomace as support and carbon source in solid-state fermentation column-tray bioreator. Biochem. Eng. J. **65**, 90–95 (2012)
42. Jayani, R.S., Saxena, S., Gupta, R.: Microbial pectinolytic enzymes: a review. Process Biochem. **10**(9), 2931–2944 (2005)
43. Uenojo, M., Pastore, G.M.: Pectinases: Aplicações industriais e perspectivas. Química Nova **30**(2), 388–394 (2007)
44. Koblitz, M.G.B.: Food Biochemistry, Theory and Practical Applications, p. 242. Guanabara Koogan, Rio de Janeiro (2008)
45. Lara-Marquez, A., Zaval-Panamo, M.G., Lopez-Romero, E., Camacho, H.C.: Biotechnological potential of pectinolytic complexes of fungi. Biotechnol. Lett. **33**, 859–868 (2011)
46. Antier, P., Minjares, A., Roussos, S., Raimbault, M., Viniegra-Gonzalez, G.: Pectinase-hyperproducing mutants of Aspergillus niger C28B25 for solid-state fermentation of coffee pulp. Enzyme Microbial. Technol. **15**, 254–250 (1993)
47. Brazil's Ministry of Health National Health Surveillance Agency: Physicochemical Methods for Food Analysis, p. 1017. MS Publisher, Brasília (2005)
48. Rangana, S.: Manual of analysis of fruit and vegetable products, p. 634. Tata McGraw Hill Publishing Company, New Delhi (1979)
49. Miller, G.L.: Use of dinitrosalicylic AID reagent for determination of reducing sugars. Analitica Chem. **31**, 426–428 (1959)
50. le Poidevin, N., Robinson, L.A.: Methods Used in the Diagnosis Of Leaf Plantations Booker Group in British Guiana: Sampling and Analysis Technique, 21st edn, pp. 3–11. Fertilité, Paris (1964)
51. Couri, S., Farias, A.X.: Genetic manipulation of *Aspergillus niger* for increased synthesis of pectinolytic enzymes. Revista de Microbiologia **26**(4), 314–317 (1995)
52. Fontana, R.C., Salvador, S., Silveira, M.M.: Influence of pectin and glucose on growth and polygalacturonase production by *Aspergillus niger* in solid-state cultivation. J. Ind. Microbiol. Biotechnol. **32**, 371–377 (2005)

53. Munhoz, C.L., Argandoña, E.J.S., Júnior, S.S.M.: Physical and chemical properties of flour obtained from guavas CV Pedro Sato. In: 10th Brazilian Congresso of Tropical Fruits, 54th Annual Meeting of the International American Society for Tropical Horticulture, Vitória-Espírito Santo 12 a 17 de Outubro de (2008)
54. Malvessi, E., Silveiera, M.M.: Influence of medium composition and pH on the production of polygalacturonase by Aspergillus oryzae. Braz. Arch. Biol. Techonol. **47**(5), 693–702 (2004)

Drying of Fruits Pieces in Fixed and Spouted Bed

Odelsia Leonor Sanchez de Alsina, Marcello Maia de Almeida, José Maria da Silva and Luciano Fernandes Monteiro

Abstract Fruit culture is one of the most diversified and important agro industrial activities in Brazil. As long as the fruit production increases and exportation grows, it is necessary to reduce losses along the production chain. Spouted and fluidized bed driers, due the high rates of mass and heat transfer, could be an alternative in order to improve the product quality. This work is concerned with drying of fruits in cone-cylindrical spouted bed, using guava pieces as a case study. Fluid dynamic studies in spouted bed and fluidized bed indicate the feasibility of using this technique for drying pieces of fruits. The diffusional model proved to be satisfactory to describe the kinetics of drying in spouted and fluidized bed. The performance of the spouted bed dryer was analyzed based on final moisture content, the loss of vitamin C and rehydration capability. It was found that, in a general way, the combined fixed bed/spouted bed process presented good performance.

Keywords Spouted bed · Fluidized bed · Kinetics of drying · Moisture content · Fruit

O. L. S. de Alsina (✉)
Institute of Technology and Research, Tiradentes University, Murilo Dantas 300, Aracaju, SE, Brazil
e-mail: odelsia@uol.com.br

M. M. de Almeida
State University of Paraiba, Campus Bodocongó, Campina Grande, PB, Brazil
e-mail: marcello_maia2000@yahoo.com.br

J. M. da Silva
Federal University of Campina Grande, Campus I, Campina Grande, PB, Brazil
e-mail: Josemariasilva2@hotmail.com

L. F. Monteiro
Federal University of Sergipe, Campus, São Cristóvão, SE, Brazil
e-mail: lucianofm2007@gmail.com

J. M. P. Q. Delgado and A. G. Barbosa de Lima (eds.), *Transport Phenomena and Drying of Solids and Particulate Materials*, Advanced Structured Materials 48,
DOI: 10.1007/978-3-319-04054-7_7,

1 Introduction

The world production of fruits has shown continued growth. According FAO data, in the period 89/91 it was 420.0 million tons, exceeded the 500.0 million tons in 1996 and in 2009, 724.5 million tons. The production of 728.4 million tons in 2010 is higher by 0.5 % compared to the previous year.

According the Brazilian Fruit Institute (IBRAF), the Brazilian production of fruit in 2013 was estimated in 43 million tons, with a slight increase of 1.65 % compared to the volume produced in 2012. Brazil is in third place in the ranking of the world's largest producers of fruits, most of them exported, behind China and India. Although much of the fruit production is exported and the domestic market is growing rapidly, a large part of the harvest is lost, especially in the case of fruits with perishable nature and fragile structure. It is estimated that in Brazil up to 40 % of fruit production is wasted.

Because they are highly perishable, high post-harvest losses of fruits causes a significant reduction in economic gains along the production chain and penalize, especially the low-income producers [1].

Among other methods of preservation, drying appears to be an attractive alternative, given the good prospects of the dried fruit world market. With the increasing demand for natural products, global consumption of dried fruit has grown in recent years. According to the Food and Agriculture Organization (FAO), 2003–2004, the volume of dried fruits exported in the world, increase 27 % from 297.2 to 376.5 tons. Brazilian exports did not exceed 30 tons [2]. Despite being a strategy for agribusiness, add value to the final quality of the dried product will depend on the quality of raw material used and especially on the type of process and operational conditions in the drying step. In preserving fruits for extending its shelf life, minimizing loss, and preventing growth of mold and microbes, it is important that the denaturation or damage of the active ingredients by the preservation methods is minimized as well [3]. As such, proper selection of dryer for the preservation of fruits is very important and vital to the process economics [4].

The good performance and product quality encourage drying of fruits and vegetables in fluidized or spouted bed. Spite the advantages of fluidized and spouted bed drying of large, deformable particles, which shrink with the water loss, is not an easy operation. The rapid variation of the physical properties, density, size, and shape as a function of the moisture content reduction can lead to instabilities, including changes in regime and spout termination [5, 6].

Despite this, stable spouting is possible in a range of operating regimes. In order to maintain this condition, a robust control of the air input velocity should be required. The knowledge of these properties and regime changes as a function of the moisture content along the time is a necessary tool to implement this kind of control, or at least to prevent the spouting instabilities [6].

Literature about drying of fruits and vegetables in fluidized or spouted bed is scarce. Combined methods, use of inerts or mechanical agitation is generally adopted to facilitate the operation. In a survey covering the last 15 years, it was found several articles about drying of fruits and vegetables in fluidized and spouted beds: Drying of sliced food products by centrifugal fluidized bed [7]; combined microwave and spouted bed to dry frozen blueberries [8]. Green beans, potatoes and peas in fluidized bed [9] drying of carrots in an inert medium fluidized bed [10] or in a mechanical agitated fluidized bed [11]. In addition, to dry carrots a spout-fluidized bed [12]. More recently [13] a solar assisted spouted bed was used to dry peas. A number of articles can be found about the drying of fruit pulps in spouted bed with inerts [14, 15].

This chapter shows the feasibility of drying fruit pieces in spouted bed, since combined with pre-drying in fixed bed. The results presented by a case study of the drying of guava show that it is possible to get a high quality product provided that the operating conditions in the spouted bed were controlled in order to maintain the spouting stability. Most of the results presented here are part of the Ph.D. theses of Almeida [16] and Silva [17].

2 Guava

The guava, native of Tropical and Subtropical America, belongs to the genus Psidium L, the Myrataceae family, which comprises more than 100 species of trees and shrubs, many of which produce edible fruit. The species *Psidium guajava L.*, whose varieties are red and white guava, has a greater commercial interest (Figs. 1 and 2).

In Brazil, the harvest period occurs from January to April, with the exception of the Para State, where the harvest is from May to August and in the Paraná State, from September to March. The fruit is ovoid, very aromatic, yellowish-green when ripe and has abundant white or red flesh.

The guava trees bear fruit in abundance, but an important part is lost due to the perishable nature of fruits and the short harvest period. Brazil is one of the largest producers of guava in the world, together with Mexico, Pakistan, Colombia, Egypt, South Africa and India, with a production relatively stable, ranging from 316,301 tons in 2007 to 328,255 tons in 2008, an area of 15,012 ha, mainly concentrated in the Southeast and Northeast [18]. As mentioned, in Brazil up to 40 % of fruit production is wasted. There are several processes of preservation and industrialization to make better use of the crop, but, until now, unlike other fruits, little is known about the production of dried guava in industrial scale. The dried guava could be consumed as such, rehydrated, or used in the food industry as raw material for products such as jams, jellies and juices or to develop new value added products.

2.1 Physicochemical Characteristics of Guava

Guava is a good source of ascorbic acid [19]; its content varies between 100 and 169 mg/100 g depending on the variety of the fruit. It has a high content of

Fig. 1 *Red* guava whole and without seeds

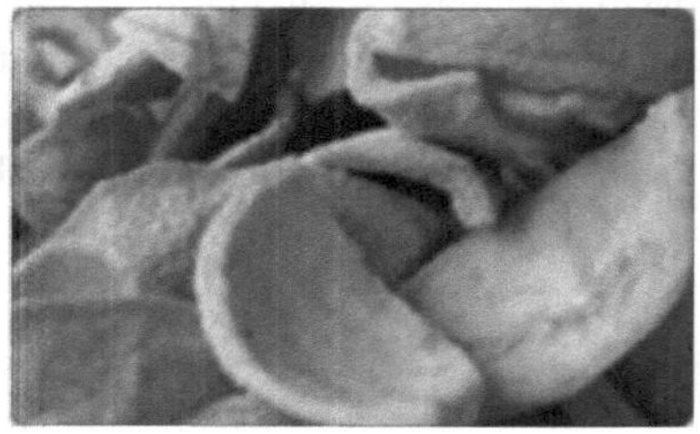

Fig. 2 Dried flesh of *white* guava

riboflavin, vitamin A activity equivalent to 1,650 units/100 g and iron from 0.2 to 0.9 mg/100 g and its energy values 62–65 Cal/100 g [20] (Table 1).

2.2 Sample Preparation

The raw material used in the experiments was red guava (*Psidium guajava L.*), variety Paluma marketed in Campina Grande—PB, Brazil. The fruits were visually selected for its consistency, maturity, color and without physical damage. Guavas were peeled, the seeds removed and the flesh cut into cubic pieces, by means of an industrial cutter as shown in Fig. 3. The mean equivalent diameter and density of the guava pieces in the form of rounded cubes were 0.011 m and 900–970 kg/m^3, respectively. After pre—drying in fixed bed the average values of equivalent diameter and density were respectively, 6.5×10^{-3} m and 870–940 kg/m^3. The fresh moisture content of the guava flesh ranged from 4.86 to 5.79 on a dry basis and between 0.6 and 1.5 after pre-drying. The pieces had an Archimedes number of the order of 3.0×10^7 and in the range of 0.6×10^7–1.0×10^7 after pre-drying.

2.3 Changes in Size and Shape Along Drying

Deformable particles such as fruits and vegetables shrink and changes in size with the water loss. The variation of the physical properties, density, size, and shape as a function of the moisture content is a disadvantage for the processing of this

Table 1 Proximate chemical composition and physical characterization of guava Paluma variety (from [20])

Chemical composition (%)	Physical characterization
Moisture (wb) 87.3 ± 0.48	Weight (g) 152.0 ± 7.8
Ashes 0.54 ± 0.03	Diameter (cm) 6.7 ± 0.2
Proteins 1.42 ± 0.06	Length (cm) 7.8 ± 0.4
Lipids 0.31 ± 0.05	Flesh thickness (cm) 1.0 ± 0.1
Total sugars 2.80 ± 0.31	Soluble solids (°Brix) 7.2 ± 0.3
Fibers 6.91 ± 0.00	pH 3.8 ± 0.0
Acidity 0.76 ± 0.01	Water activity 0.99 ± 0.002

Fig. 3 Preparation of guava pieces (From [17])

category of materials, because the continuous changes in transport properties [10]. Several authors studied the effect of shrinkage on the drying process for carrots, apples, potatoes, garlic, tomato, West Indian Cherry, among others [10, 21–23]. It is usually reported that shrinkage of material during drying is a linear function of linear dimension and moisture content [24, 25]. Other authors considered the quadratic relation, equivalent to an area shrinkage, with the variation of $(r/r_o)^2$ as a linear function of the moisture content [23, 26]. In spite of the strong relation between moisture content and physical properties, the volume reduction does not always present a direct correlation with the amount of water evaporated. Then, the shrinkage is not only a function of moisture content, but also depend on the operating conditions and sample geometry [22]. The shrinkage behavior is different for various systems, dependent on the material type, the cell and tissue structure, and operating conditions [10]. This behavior was studied for guava pieces at several drying conditions along fixed, fluidized and spouted bed [6, 16, 27]. An overview of the obtained results is presented below.

2.3.1 Density

Within the perspective of spouted bed drying, the particle density is an important parameter together the size to analyze the spout stability and fluid dynamics,

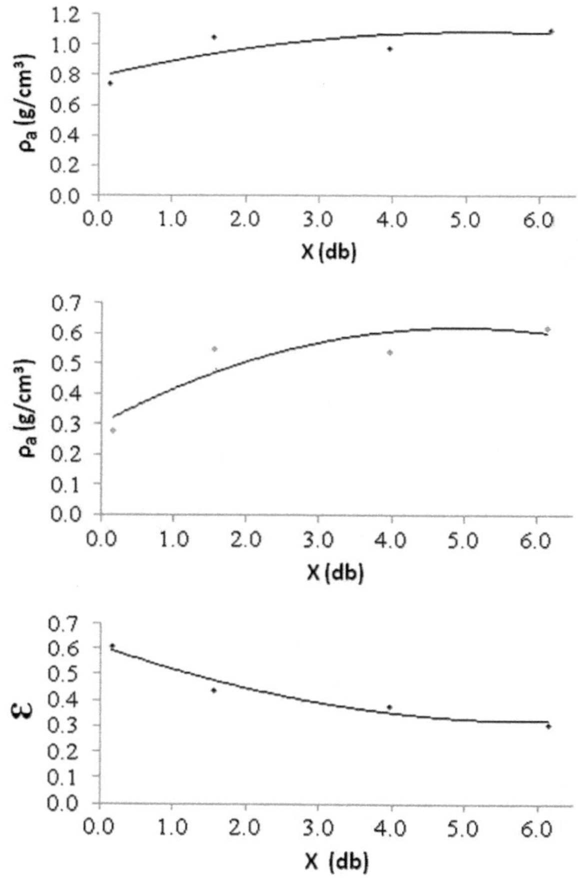

Fig. 4 Bed Bulk density (**a**), particle density (**b**) and bed porosity (**c**) of cubic pieces of guava dried in an oven at 60 °C (From [17])

because both properties affect the Archimedes number. The effect of moisture on the density of the product was obtained by drying the guava and measuring the properties by picnometry at different levels of moisture. Figure 4 shows the results for pieces dried in an oven with air circulation at 60 °C.

In order to verify the influence of the process, it was also determined the variation of the properties with the changes in moisture during the drying in a tray dryer with air circulation at 50, 60 and 70 °C and during the spouted bed drying at 60 °C. From these results, showed in Fig. 5 it can be seen that the density suffers significant variation in high moisture levels with different operating conditions of temperature. In this region, the amount of water removed is equivalent to the variation of reducing the size of cubic pieces.

It can be seen from Figs. 4 and 5, that the density reaches a maximum at around 2.5 moisture conditions on a dry basis. It is observed from both Figures that the experimental data of density obtained during the drying experiments of cubic

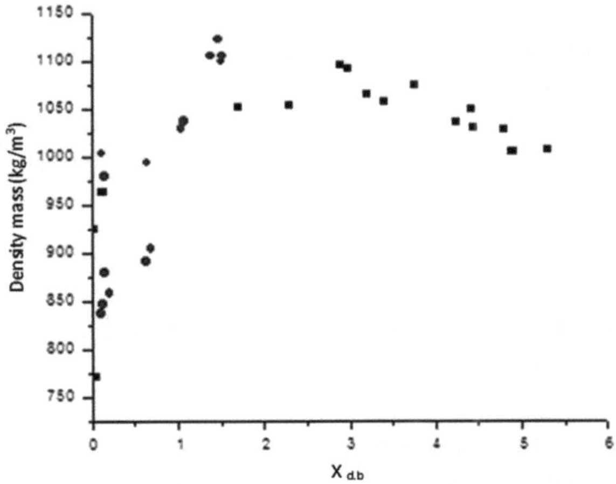

Fig. 5 Particle density as function of moisture content of guava pieces dried in (*square*) Tray dryer at 50 °C, 60 °C and 70 °C; (*dot*) spouted bed dryer at 60 °C (From [16])

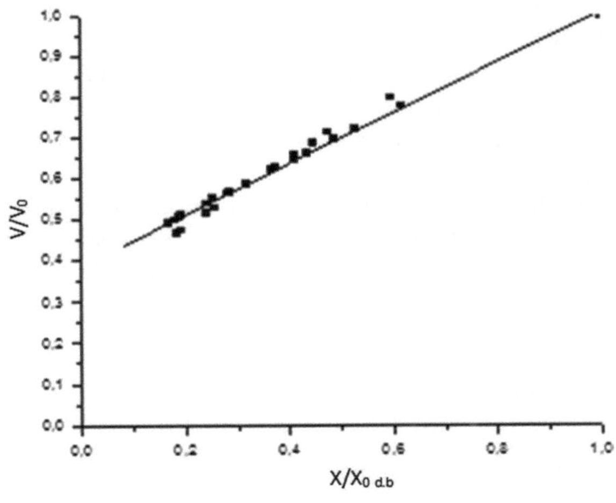

Fig. 6 Shrinkage of guava pieces dried in stove at 40 °C, 60 °C and 70 °C. $X_o = 0.6$ and 1.5

pieces of guava in spouted bed show a similar trend that was found for fixed bed at different temperatures.

2.3.2 Volumetric Shrinking

The volume of the equivalent sphere was calculated by liquid picnometry. The volumetric shrinkage, represented in normalized form V/V_o versus X/X_o showed a linear response throughout the drying in stove as shown in Fig. 6. Therefore we used a linear model proposed by [28], with constant coefficient of shrinkage. The model, Eq. (1), fits fairly well the experimental values obtained, as shown in the

Fig. 6 and the coefficient of determination R^2 was 0.9966. The volumetric shrinkage coefficient, β, equal to 0.6224, was not influenced by temperature in the studied range.

$$\frac{V}{V_o} = 0.38621 + 0.6224\frac{X}{X_o} \tag{1}$$

3 Fluid Dynamics of the Spouted Bed Drying of Guava Pieces

Fluid dynamic conditions are important in the study of drying kinetics, in particular because the bed fluid dynamic behavior depends on properties that are moisture content dependents. Despite a large number of articles about behavior in dynamically active beds, studies involving deformable particles are scarce [5, 6, 9, 29, 30].

To avoid the fluid dynamic instabilities and product degradation a mechanically agitated fluidized bed dryer was used for drying carrots [11]. Another option for sliced food products was by fluidization and drying in a centrifugal fluidized bed dryer [7].

In the case study of the present work, it was observed that the rounded cubes of guava belong to the class of spoutable particles according Geldart as we can see in Fig. 7, where the domain of the particles is in the D region.

3.1 Experimental Device

The tests of fluid dynamics and drying of guava pieces in spouted bed were performed in the experimental device shown in the Fig. 8. The Figure shows a picture of the system, which consists of a column in acrylic, air blower, electrical heat exchanger, valves to control the airflow rate and a data acquisition system. The column has orifices for the pressure, temperature and air humidity probes. An anemometer at the top of the column measured the air velocity. The tests were performed with two cone-cylindrical columns with the following dimensions:

Column A: cylindrical part with height of 1.0 and 0.108 m diameter. The conical section at the bottom of the column has a height of 0.09 m. The air inlet orifice has a diameter of 0.028 m.

Column B: Cylindrical part with height of 1.0 and 0.25 m diameter. The conical section at the bottom of the column with 0.25 m in height and air inlet hole of 0.05 m.

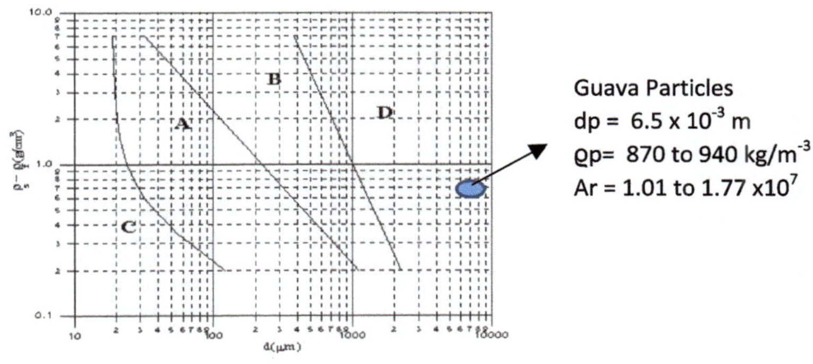

Fig. 7 Domain of the guava particles in Geldart's map

Fig. 8 Picture of the experimental device (From [17])

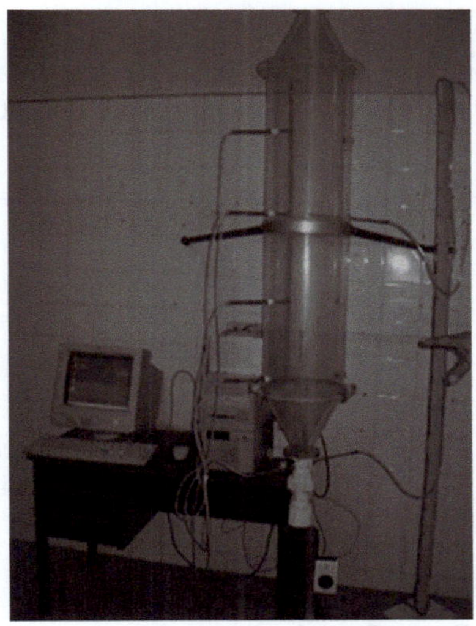

3.2 Flow Regimes

Due the continuous change in the physical properties of particles, the characteristic curves are unusual and do not follow the expected trend for spouted or fluidized bed. The pieces of guava changed the moisture content, size and density during each experimental run. Figure 9 illustrate the evolution of the properties, normalized with their respective initial value for one experiment. Similar behavior was observed in other experiments.

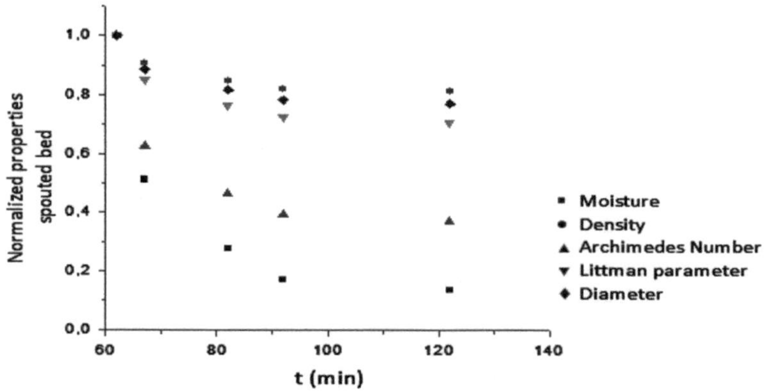

Fig. 9 Evolution of the normalized properties during an experimental run. Spouted bed in column A (From [16])

The Littman parameter A is defined by Eq. (2) and the Archimedes number by Eq. (3). Both parameters are important in analyzing the spout regime and stability.

$$A = \frac{\rho_g v_t v_{mf}}{(\rho_s - \rho_g) g D_i} \tag{2}$$

$$Ar = \frac{d_p^3 g \rho_g (\rho_s - \rho_g)}{\mu^2} \tag{3}$$

where:

V_t Terminal velocity of the particle (m)
V_{mf} Minimum fluidization velocity (m/s)
ρ_g Air density (kg/m^3)
ρ_s Particle density (kg/m^3)
D_i Input orifice diameter (m)
D_p Particle diameter (m)
μ Air viscosity (Pa s)

Different ranges of parameter A, Littman parameter [31], can identify the mechanism of the spout termination in the cone-cylindrical geometry [32]. This parameter represents the ratio between the minimum energy required to form the spout and that required to keep it along the bed under conditions of minimum spouting, at the maximal spoutable height. This was important for the experiments in column A where the ratio d_p/D_c was 0.06, greater than in column B, where the d_p/D_c was 0.026. The spouting in column A corresponds to that of large particles and in this kind of spouted beds there are structural changes leading to instability, regime change or collapse of the fountain. For this column, the initial conditions for the Littman and Archimedes numbers in the different experiments were very close, between 0.11 and 0.15 and 1.01×10^7–1.77×10^7, respectively. The spout collapse as termination mechanism occurs for fine particles (dp < 1 mm) or, in the case of large particles, for values of A < 0.014 [32]. However, there is experimental evidence that this is the termination mechanism in the case of guava pieces in column A where dp = 10 mm; 4, Littman parameter = 0.11–0.15). The observed termination slugging-spout suggests that the primary mechanism for termination corresponds to the beginning of bubble formation at the inlet orifice rather than fluidization at the top of the annulus. This was confirmed by visual observation and the fact that $v_{ms} < V_{mf.}$

Figure 10 illustrates the different regimes observed in an experimental run along the drying of guava pieces.

At the start, the material with high moisture content remains packed in a fixed bed Fig. 10a. Three spout regimes were observed: a stable spouting regime at high moisture content (b), a fast-fluidized bed regime slugging at intermediate moisture values (c), and a spouted bed (d). Under the conditions of Fig. 10d, it can be observed an elevated height of the fountain. During the spouting regime, the fountain height increases as the particles dry up. In addition, a pneumatic transport regime was observed in some experiments at very low moisture content.

The regime transitions found operating the column A with guava pieces were mapped as a function of the moisture content. Figure 11 shows the four regions corresponding to different moisture levels. For moisture content greater than 0.8 the drier works as a fixed bed unless the air velocity is greater than 2.5–3 times the minimum spouting velocity. As the material dries the bed will working successively as fluidization, spouting and slugging.

In the column B, with a diameter of 0.25 m and particles of the same diameter than in the experiments performed in column A, the spout regime was stable for guava pieces with moisture content lower than 1.5. Above this value, the wet particles remain packed in fixed bed or slightly expanded until dried enough; then the fountain emerged and a stable spout was established.

4 Drying Kinetics

From the results obtained in the fluid dynamic studies it was decided to hold the drying in combined fixed bed-spouted bed, so that the initial moisture content of guava pieces to finished the drying in spouted bed was less or equal to 1.5(d.b).

Figure 12 shows the drying curves in combined fixed bed, first 60 min, and spouted bed in the finish drying step. Despite the high initial moisture content of the fresh fruit, in both fixed bed and spouted bed, drying was in the period of decreasing rate.

A simple Fickian model with constant effective diffusion coefficient can describe the drying kinetics of agricultural products and tropical fruits. However, it is common to find kinetic behavior that deviates from this simple model, which is not adequate to describe the operation completely. Some authors attribute the

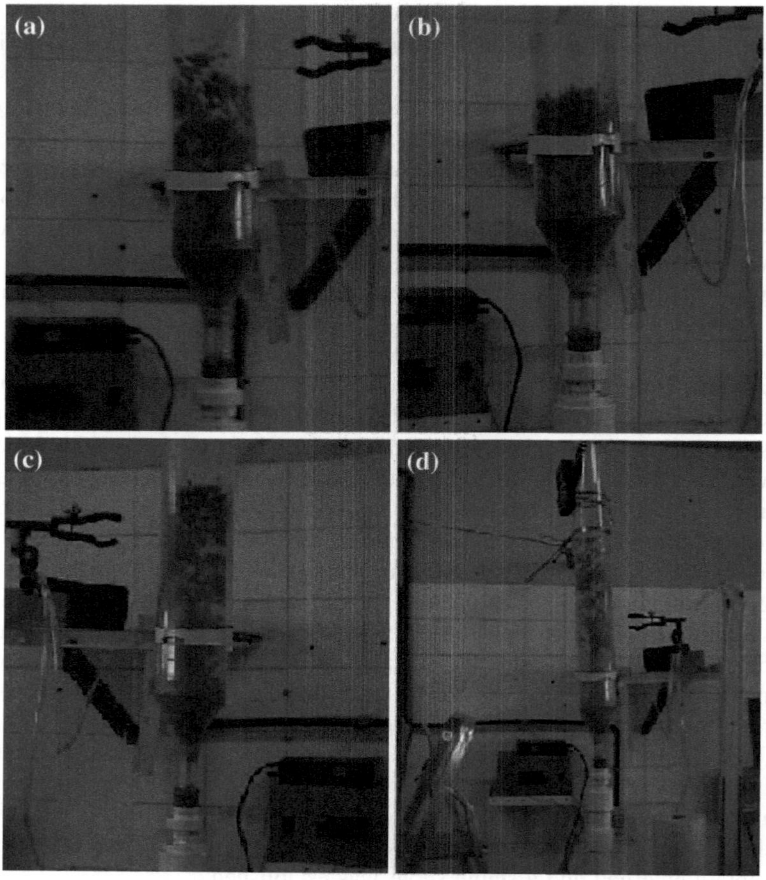

Fig. 10 Flow regimes in the spouted bed column A **a** fixed bed; **b** expanded bed with guava circulation at the *top* of the annulus; **c** fluidization with slugging; **d** spouted bed (From [16])

Fig. 11 Map of flow regimes in the spouted bed, column A

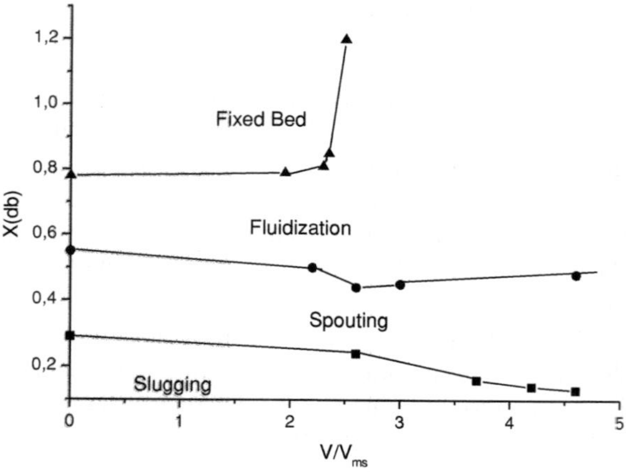

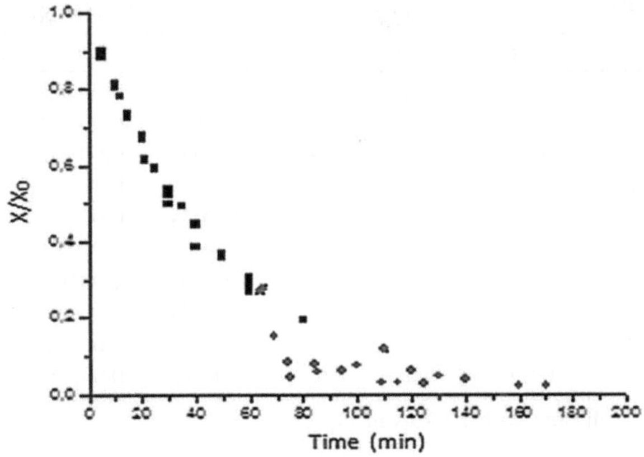

Fig. 12 Combined fixed-spouted bed drying curves

deviation from Fickian model to the incidence of other non-diffusional mechanisms. One reason for the failure of the model, especially in case of fruits, is the effect of shrinkage, which is not considered in most of the applications [23]. In this work, the shrinkage was included by a hypothesis of pseudo-constant volume. In this simplified approach, the shrinkage is considered by including a moisture dependent dimension in the constant volume Fick's solution. The rounded guava pieces were approximate by a same volume sphere with equivalent diameter moisture dependent. The Fick's solution for the constant volume sphere is [33], with initial and boundary conditions:

$$X(r,0) = X_o \tag{4a}$$

$$X(R_p,t) = X_\infty \tag{4b}$$

$$\frac{\partial X}{\partial t}(0,t) = 0 \tag{4c}$$

$$X^* = \frac{6}{\pi^2}\sum_{n=1}^{\infty}\frac{1}{n^2}\exp\left(-\frac{n^2\pi^2 D_{ef} t}{R_p^2}\right) \tag{5}$$

$$X^* = \frac{X - X_\infty}{X_o - X_\infty} \tag{6}$$

where X is the mean moisture content at time t, X_o is the initial moisture content (db), X_∞ is the equilibrium moisture content (db), D_{ef} is the effective diffusion coefficient of water in the solid (m^2/s), R_p is the particle radius at time t (m), t is the time (s) and R_p is the instantaneous radius calculated as a function of moisture according to the volumetric shrinking, Eq. (1).

The Fickian model, Eq. (5), was adjusted to the experimental drying curves with $t/R_p^2(X)$ as independent variable. The theoretical value $6/\pi^2 = 0.6079$ was

Table 2 Kinetics parameters obtained from the Fickian model (From [34])

Run	H_o (m)	X_o	v_o/v_{ms}	C	D_{ef} (cm^2/min)	R^2	Standard deviation
1	0.09	0.6	2.0	0.7015	0.00063	0.99	0.00002
2	0.12	0.6	1.07	0.6875	0.00049	0.97	0.00009
3	0.09	1.5	1.10	0.6667	0.00094	0.96	0.00015
4	0.12	1.5	1.00	0.7009	0.00168	0.99	0.00026
5	0.09	0.6	2.50	0.7034	0.0006	0.99	0.00002
6	0.12	0.6	1.30	0.7020	0.00053	0.99	0.00005
7	0.09	1.5	1.40	0.6765	0.00101	0.96	0.00019
8	0.12	1.5	1.25	0.6884	0.00081	0.98	0.0001
9	0.105	1.05	1.88	0.6420	0.00041	0.94	0.00009
1[a]	0.09	0.6	2.0	0.6613	0.00059	0.99	0.00002

[a] Adjusted with 7 terms in Eq. (5)

replaced by and adjustable parameter C. Four terms of the series were considered and the D_{ef} and C parameters determined by nonlinear regression using Levenberg Marquardt method.

Table 2 shows the fitting results of the parameter C and the effective diffusion coefficient. It is found that the values for D_{ef} vary between 4.0×10^{-4} and 1.7×10^{-3} cm^2 min^{-1}.

In most experiments, the diffusion sphere model considering the volumetric shrinkage showed a good fit to the experimental data, with significant correlation coefficients of around 99 %, which allowed an estimate of the effective diffusivity from the model.

From Table 2 it appears that there is no influence of the air velocity on the D_{ef}. This suggests that the consideration of negligible external resistance and internal diffusion control is a realistic assumption. Notice that the parameter C of the fitted curve is close to the theoretical value of 0.6079 except in three experiments. Comparing the results 1 and 1^*, where the former corresponds to four terms in the series whereas in the 1^* the same data were modeled with 7 terms, it is observed that there is a natural tendency to the parameter C approach to its theoretical value as more terms in the series were included.

Figure 13 shows the drying kinetics, experimental curves and model. From the Fig. 13 and Table 2 it can be seen that, in general, the model presented a good fitting to the experimental data, with determination coefficients above 0.94 and thus, reliable estimates of the diffusion coefficient for the drying of guava pieces in spouted bed.

5 Product Quality

Table 3 presents the results of the analysis of the degradation of vitamin C in guava, dry basis, during the pre-drying of guava pieces in fixed bed followed by drying in spouted bed in column A. It is found that for approximately equal values

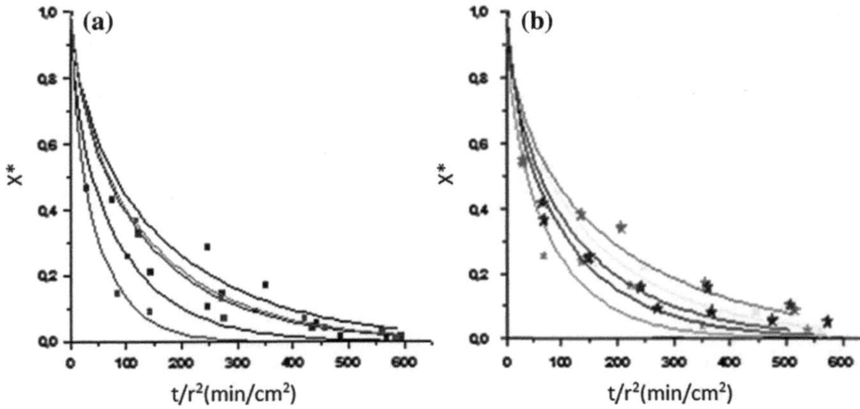

Fig. 13 Drying kinetics in spouted bed. Column A: **a** experiments 1–5; **b** experiments 6–10;— Fick's Diffusional model including shrinkage (From [34])

Table 3 Loss of vitamin C in the drying process combined fixed bed/spouted bed, Column A (From [16])

Experimental conditions	a	b	c	FB loss (%)	SB loss (%)
	11.150	8.254	5.532	25.97	32.97
$H_o = 0.12$ m; $X_o = 0.6$; $V/V_{jm} = 1.07$	6.208	4.886	4.825	21.29	1.25
	_	5.301	4.758	_	10.24
$H_o = 0.12$ m; $X_o = 1.5$; $V/V_{jm} = 1.0$	6.855	3.441	3.415	49.80	0.76
$H_o = 0.09$ m; $X_o = 0.6$; $V/V_{jm} = 2.5$	11.816	9.057	8.254	23.34	8.86
$H_o = 0.12$ m; $X_o = 0.6$; $V/V_{jm} = 1.3$	_	7.947	7.484	_	5.83
	7.280	5.384	4.662	26.04	13.41
$H_o = 0.12$ m; $X_o = 1.5$; $V/V_{jm} = 1.25$	6.855	3.283	3.176	52.11	3.26
	–	5.940	5.170	–	13.00
$H_o = 0.105$ m; $X_o = 1.05$; $V/V_{jm} = 1.88$	–	6.950	6.08	–	12.50

[a] Initial value (% db)
[b] After drying in fixed bed (% db)
[c] After drying in spouted bed (% db)

of air velocity to minimum spouting velocity, V/V_{ms} ratio, the losses of vitamin C are more pronounced in the experiments with the lowest load of guava and higher initial moisture. Comparing the experiments 2 and 5, it can be seen that the effect of V/V_{ms} is significant for the same conditions of H_o and X_o. The Vitamin C losses are more pronounced at the lowest conditions for the initial values of the independent variables. Moreover, this behavior is contrary to the conditions of most of the control variables of the process. It is observed that the independent variables, initial bed height and initial moisture content of the product, have a negative effect on vitamin C, the higher the load and the higher the initial moisture, smaller vitamin losses.

Table 4 Moisture, water activity, vitamin C and percentage of rehydration of the product as a function of the initial moisture and drying temperature in spouted bed, Column B

T (°C)	X_o (d.b)	X_f (d.b)	a_w	VC (mg/100 g)	Rh (%)
50.0	0.60	0.34	0.48	142.00	65.89
50.0	1.50	0.19	0.33	84.50	68.41
60.0	0.60	0.33	0.45	126.60	66.33
60.0	1.50	0.25	0.39	97.50	67.71
47.9	1.05	0.20	0.34	87.60	68.17
62.1	1.05	0.37	0.56	156.20	64.13
55.0	0.41	0.34	0.53	142.00	64.24
55.0	1.69	0.22	0.38	89.10	67.92
55.0	1.05	0.28	0.43	112.60	67.52
55.0	1.05	0.27	0.39	104.10	67.72
55.0	1.05	0.25	0.39	97.50	67.79

The results shown in Table 3 were obtained in column A with constant working temperature, around 50 °C. In experiments performed in the column B, the effect of temperature was also investigated. It was found that the initial moisture X_o affects the losses of vitamin C. Table 4 presents the results obtained in the column B.

According to Table 4, it is observed that, as expected, the highest remainder amount of vitamin C, (156.2 mg/100 g), was obtained at the lowest drying temperature, 47.9 °C. The fresh guava pulp had a mean value of 211, 14 mg ascorbic acid/100 g. Like most fruits, the variability of the composition of guava is large. According [35] guava present ascorbic acid content that can vary between 55 and 1044 mg ascorbic acid per 100 g pulp. From the regression analysis, the final average value for the pieces of guavas dried in spouted bed was 104.73 mg/100 g. Thus, the vitamin C amounts remaining after drying in spouted bed can be considered good without considerable losses.

The final moisture and water activity are extremely low, ensuring a safe storage and indicating that it could be used a shorter drying time in the finish step.

The Fig. 14 shows the kinetics of rehydration of guava pieces after drying in spouted bed at three experimental conditions. It was found that the kinetic behavior of rehydration was very similar for the three conditions of the spouted bed drying, varying X_o and T.

It is known that rehydration is a complex process based on a reconstruction of the properties of certain dry material in contact with a liquid phase. The rate and amount of water absorbed in this process are dependent on each material and previous process. These authors studied the kinetics of rehydration of banana, pumpkin, apple and other vegetables and found that the rehydration could vary between one and four times in depending of the type of material. According to them, this behavior is due to the hysteresis effect in the drying process by disruption of cellular structures. In the case of guava pieces dried in combined fixed/spouted bed, the product absorbs water to approximately 65 %, reaching this in

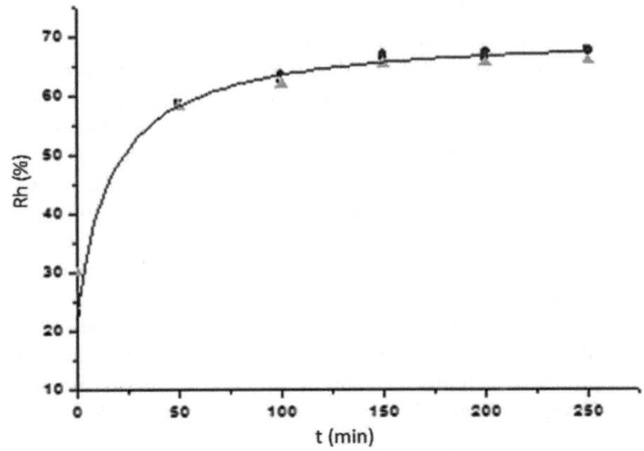

Fig. 14 Kinetics of rehydration of the dried guava pieces. T = 50 °C $X_i = 0.6$ $X_f = 0.25$; T = 55 °C $X_i = 1.07$ $X_f = 0.28$; T = 60 °C $X_i = 1.54$ $X_f = 0.34$

2 h. Structural changes during drying would be responsible for the partial rehydration of guava just up to 65 %.

The foregoing results indicate that the product of combined drying has good properties, maintaining the nutritional quality.

6 Final Remarks

The study of drying guava in combined fixed bed-spouted bed pointed to the feasibility of this technique for the processing of fruit pieces. Special care should be taken in the design of the equipment considering the geometry, especially the particle diameter to column diameter ratio, as well as the choice of appropriate operating conditions to ensure good stability of the spouting regime for a correct scaling up. It was observed that the dehydrated product obtained in this case study, presented good quality nutritional and organoleptic properties as well a final moisture content proper to storage and rehydration capability. These results allow us to expect good outcomes with other fruits similar to Guava in their textural characteristics.

Acknowledgments The authors would like to acknowledge CNPq, National Scientific and Technologic Development Council of Brazil for the scholarships and financial support.

References

1. Arruda, M.C., Jacomino, A.P., Sarantópoulos, C.I.G.L., Moretti, C.I.: Quality of minimally processed melon stored under passive modified atmosphere. Hortic. Bras. **21**(4), (2003) (In Portuguese)
2. Spers, E.E., Begiato, G.F., Castro, L.T., Neves, M.F.: Dried fruit market. In Agro ANALYSIS. Revista de agronegócios da Fundação Getúlio Vargas. (Journal on line,

December 2008) http://agroanalysis.com.br/materia_detalhe.php?idMateria=567. Accesed in 10 April 2014 (in Portuguese)
3. Santos, P.H.S., Silva, M.A.: Retention of vitamin C in drying processes of fruits and vegetables—A review. Drying Technol. **26**, 1421–1437 (2008)
4. Fernandes, F.A.N., Rodrigues, S., Law, C.L., Mujumdar, A.S.: Drying of exotic tropical fruits: a comprehensive review. Food Bioprocess Technol. **4**, 163–185 (2011)
5. Almeida, M.M., Alsina, O.L.S., Silva, O.S, Costa, M.G.: Drying of cubic guava pieces in spouted bed. Part I: fluid dynamic study. In: 14th International Drying Symposium. IDS2004, São Paulo. Proceedings of 14th International Drying Symposium. IDS2004. Campinas: Ourograf. Gráfica e editora, vol. C, pp. 1784–1791 (2004)
6. Almeida, M.M., Silva, O.S., Alsina, O.L.S.: Fluid-dynamic study of deformable materials in spouted bed dryer. Drying Technol. **24**(4), 499–508 (2006)
7. Shi, M.H., Hao, Y.L., Din Y.T.: An experimental investigation on the drying of sliced food products in centrifugal fluidized bed I. J. Therm. Sci. **7**(3), 181–185 (1998)
8. Feng, H., Tang, J., Mattinson, D.S., Fellma, J.K.: Microwave and spouted bed drying of frozen blueberries: the effect of drying and pretreatment methods on physical properties and retention of flavor volatiles. J. Food Process. Preserv. **23**, 463–479 (1999)
9. Senadeera, Wijitha, Bhandari, Bhesh R., Young, Gordon, Wijesinghe, Bandu: Influence of shapes of selected vegetable materials on drying kinetics during fluidized bed drying. J. Food Eng. **58**, 277–283 (2003)
10. Hatamipour, M.S., Mowla, D.: Shrinkage of carrots during drying in an inert medium fluidized bed M.S. J. Food Eng. **55**, 247–252 (2002)
11. Reyes, P.I., Alvarez, Marquardt F.H.: Drying of carrots in a fluidized bed I effects of drying conditions and modelling. Drying Technol. **20**(7), 1463–1483 (2002)
12. Zielinska, Magdalena, Markowski, Marek: Drying behavior of carrots dried in a spout-fluidized bed dryer. Drying Technol. **25**, 261–270 (2007)
13. Sahin, Serpil, Sumnu, Gulun, Tunaboyu, Ferihan: Usage of solar-assisted spouted bed drier in drying of pea. Food Bioprod. Process. **9**, 271–278 (2013)
14. Medeiros, M.F.D., Rocha, S.C.S., Alsina, O.L.S., Jerônimo, C.E.M., Medeiros, U.K.L., da Mata, A.L.M.L.: Drying of pulps of tropical fruits in spouted bed: effect of composition on dryer performance. Drying Technol. **20**(4&5), 855–881 (2002)
15. Rocha, S.C.S., Souza, J.S., Alsina, O.L.S., Medeiros, M.F.D.: Drying of tropical fruit pulps: spouted bed process optimization as a function of pulp composition. Drying Technol. **29**, 1587–1599 (2011)
16. Almeida, M.M.: Drying of slices and diced guava pieces (Psidium guajava L.). Ph.D. Thesis, Federal University of Campina Grande, Brazil (2004) (In Portuguese)
17. Silva, J.M.: Combined drying of cubic guava pieces in fixed bed and spouted bed, Ph.D. Thesis, Federal University of Campina Grande (2011) (In Portuguese)
18. AGRIANUAL.: Yearbook of the Brazilian Agriculture, 10th edn. São Paulo, FNP Consultoria and Agroinformativos. (2010) In Portuguese
19. Sanjinez-Argandoña, E.J., Menegalli, F.C., Cunha, R.L. da., Hubinger, M.D.: Evaluation of total carotenoids and ascorbic acid in osmotic pretreated guavas during convective drying. Italian J. Food Sci. Itália. **17**(3), 305–314 (2005)
20. Pereira, L.M., Rodrigues, A.C.C., Sarantópoulos, C.I.G. de L., Junqueira, V.C.A., Cardello, H.M.A.B., Hubinger, M.D.: Shelf life of minimally processed guava packed in modified atmosphere packaging. Ciência e Tecnologia de Alimentos. Campinas. **23**(3), 427–433 (2003) (In Portuguese)
21. Kross, R.K., Mata, M.E.R.M.C., Duarte, M.E.M.: Shrinkage effect during drying of fresh pretreated tomates. Revista Brasileira de Produtos Agroindustriais **4**(2), 187–194 (2002)
22. Ratti, C.: Shrinkage during drying of foodstuffs. J. Food Eng. **23**, 91–105 (1994)
23. Alsina, O.L.S, Silva, O.S., Brasileiro, M.N.: Drying kinetics of Indian cherry. In: Proceedings of the Inter-American Drying Conference (IADC), Itu-SP, pp. 434–440 (1977)

24. Park, K.J., Tuboni, C.T., de Oliveira, R.A., Park, K.J.B.: Study of drying of Persimmon Giombo with and without shrinkage. Revista Brasileira de Produtos Agroindustriais, Campina Grande **6**(1), 69–84 (2004). In Portuguese
25. Hernandez, J.A., Pavon, G., Garcia, M.A.: Analytical solution of mass transfer equation considering shrinkage for modeling food-drying kinetics. J. Food Eng. **45**, 1–10 (2000)
26. Achanta, S., Okos, M.R., Cushman, J.H., Kessler, D.P.: Moisture transport in shrinking gels during saturated drying. AIChE J. **43**(8), 2112–2122 (1997)
27. Alsina, O.L.S., Maia, M.M., Aragão, R.F.: Study of thermal properties of guava (Psidium guajava) and cashew (Anacardium occidentalis L.). In: XV COBEQ Congresso Brasileiro de Engenharia Química, Curitiba, vol. 1, pp. 1–8 (In Portuguese) (2004)
28. Keey, R.B.: Drying of Loose and Particulate Materials. Hemisphere Publishing Corporation, Washington (1992)
29. Senadeera, W., Bhandari, B., Young, G., Wijesinghe, B.: Physical properties and fluidisation behaviour of fresh green bean particulates during fluidised bed drying. Food Bioprod. Process. (Trans IchemE) **78**, 43–47 (2000)
30. Palzer, S., Dubois, C., Gianfrancesco, A.: Generation of product structures during drying of food products. Drying Technol. **30**(1), 97–105 (2012)
31. Morgan, N.H., Littman, H.: Predicting the maximum spouted height in spouted beds of irregularly shaped particles. Ind. Eng. Chem. Fundam. **21**, 23–26 (1982)
32. Passos, M.L.A., Mujumdar, A.S.: Effect of cohesive forces on fluidized and spouted beds of wet particles. Powder Technol. **110**, 222–238 (2000)
33. Crank, J.: The mathematics of diffusion, 2nd edn. Oxford University Press, New York (1975)
34. Almeida, M.M., Silva, O.S., Alsina, O.L.S.: Kinetic study of cubic guava pieces dried in spouted bed. In: 4th Mercosur Congress on Process Systems Engineering, 2005, Rio de Janeiro. 4th Mercosur Congress on Process Systems Engineering/Enpromer Proceedings (2005)
35. Rathore, D.S.: Effect of season on the growth and chemical composition of guava (Psidium guajava L.) fruits. J. Hortic. Sci. Biotechnol. (Ashford Kent) **51**(1), 41–47 (1976)

Nutricional Enrichment of Waste from Mesquite Pods (*Prosopolis Juliflora*) Using *Saccharomyces Cerevisiae*

M. B. Muniz, F. L. H. da Silva, J. P. Gomes, S. F. M. de Santos, F. C. dos Santos Lima, H. V. Alexandre and A. S. Rocha

Abstract The mesquite is an exotic tree is a plant from the region of Piura in Peru, which is easily adapted in Northeast Brazil. Today, several studies confirm its usefulness for both animal feed and human. With the growing concern for the environment and sustainability harnessing its waste has been studied. Thus, this study was conducted with the aim of studying the isotherms as well as analyzing the effect of inoculation of *Saccharomyces cerevisiae* in the waste of mesquite pods in the protein enrichment in semisolid fermentation process. It was found that the GAB model was best represented the experimental data of the isotherms. The values of total reducing sugars were mostly inversely proportional to protein

M. B. Muniz (✉) · H. V. Alexandre · A. S. Rocha
Post-Graduate in Process Engineering, Center of Science and Technology, Federal University of Campina Grande, Campina Grande, Paraiba, Brazil
e-mail: mbmmuniz@yahoo.com

H. V. Alexandre
e-mail: hofsky@gmail.com

A. S. Rocha
e-mail: alleksandra_leka@yahoo.com.br

F. L. H. da Silva · S. F. M. de Santos
Department of Chemical Engineering, Center of Technology, Federal University of Paraiba, João Pessoa, Paraiba, Brazil
e-mail: flavioluizh@yahoo.com.br

S. F. M. de Santos
e-mail: sharlinefm@hotmail.com

J. P. Gomes
Department of Agricultural Engineering, Center of Science and Technology Federal, University of Campina Grande, Campina Grande, Paraiba, Brazil
e-mail: josivanda@gmail.com

F. C. dos Santos Lima
Science and Technology Federal Institute of Education—IFET/PE, Campus Belo Jardim, Pernambuco, Brazil
e-mail: flavia.c.7@hotmail.com

J. M. P. Q. Delgado and A. G. Barbosa de Lima (eds.), *Transport Phenomena and Drying of Solids and Particulate Materials*, Advanced Structured Materials 48,
DOI: 10.1007/978-3-319-04054-7_8,

increases. The analysis of the material obtained presented a maximum content of 26.27 % crude protein at 40 °C over a period of 24 h.

Keywords Mesquite · Residue · Fermentation · Drying

1 Introduction

The mesquite is a leguminous plant which is rich in nutritional components and found aplenty in the Northeast Region of Brazil, especially in the Cariri region of Paraíba state. With the constant growth of the horticulture agribusiness in Brazil's Northeast, the use of by-products from processing stands out as an alternative for animal feed. It is noteworthy that, despite this growth, agro-industrial by-products are still little used and may become environmental pollutants [29].

The mesquite is a leguminous tree that focuses most of its nutritional value on pods (fruits), and thereby it is a rich source of carbohydrates and proteins. Studies on the use of mesquite for several species, such as cattle, sheep, pigs, poultry, have been developed with the goal of making feasible its inclusion in feed, as well as minimizing the costs of animal production [7, 16, 17]. In order for small industries, and small and medium producers to grow and be competitive, studies bringing improvements for farmers need to be carried out, adding value to products through better use, such as the industrialization of pods and their byproducts.

The use of regionally adapted plant material is essential to improving the food supply that can replace, partially or completely, some basic components in the composition of animal feed. However, in the case of tropical fruits it is possible to observe a great waste of by-products from processing with excellent potential for use in animal nutrition, as is the case of cashew [22].

The following substrates used for obtaining protein enrichment by microorganisms stand out: banana waste, flour and solid waste from processing cassava, corn cob, sugar cane, sugar cane bagasse, apple pomace, molasses, vinasse, wheat bran and straw, chick- peas, coffee pulp, cooked rice, lemon peel, orange, waste from the manufacture of frozen acerola pulp, passion fruit, pineapple, guava, strawberry and cashew peduncle. These in Brazil represent abundant raw material of low cost [8, 11, 20, 21, 25].

The biggest issue concerning the use of protein-rich foods is their high cost. Once the producers can enrich the forages on their property with protein, the cost of food protein will decrease because there will be no transportation expenses of feed and they will be using only material produced on their farms [4, 20, 22].

The semisolid state fermentation (referred to in Portuguese as FSS) is a process in which there is growth of microorganisms on or within the solid matrix particles, where the quantity of liquid has a water activity level that can ensure the growth and metabolism of the microorganisms but which does not exceed the maximum binding capacity of the water [21].

Almeida et al. [2] analyzed the bioconversion of mandacaru for the production of bioproducts, and they found that during the semisolid fermentation of mandacaru using the fungus *Aspergillus niger*, there was an increase of 76.9 % protein in the substrate, in the time of 72 h fermentation at a temperature of 30 °C to a thickness of 1 cm.

Campos et al. [7] investigated the nutritional enrichment of cactus pear and observed that the maximum crude protein content achieved in the rotating drum bioreactor was 43.27 %, attaining a 6.44 times increase of protein using the yeast *Saccharomyces cerevisiae*. The same authors also noted that the best results for obtaining the crude protein (protein enrichment) was at the fermentation time of 8 h in trays, and at 4 h for fermentation in rotating drums.

In addition, Holanda et al. [11] studied the protein enrichment of cashew peduncle with the yeast *Saccharomyces cerevisiae* for animal feed and noted that the proportion of 10 % of yeast on the biomass from the cashew's peduncle provides a protein content above the 20 % of the fermented material. These authors concluded that the fermentation time needed for the protein conversion of the carbohydrates from the cashew's peduncle is less than 24 h.

Finally, Oliveira et al. [19] worked with bioconversion nutritional enrichment of organic residues to use in animal feed and noted that the best values of the concentration of crude protein for the waste occurred at the time of 48 h using the yeast *Saccharomyces cerevisiae*.

The present work aims at developing the protein enrichment of waste from mesquite pods, by means of the semisolid fermentation using the organism *Saccharomyces cerevisiae*, seeking to maximize the increase in protein content in the substrate.

2 Materials and Methods

This work was carried out at four laboratories, namely the Laboratory of Biochemical Engineering (referred to in Portuguese as LEB) of the Academic Unit of Chemical Engineering UAEQ/CCT/UFCG, the Laboratory of Transfer in Porous Media and Particulate Systems Academic Unit of Chemical Engineering UAEQ/CCT/UFCG, the Electrochemistry Engineering Laboratory from the Federal University of Campina Grande UAEQ/CCT/UFCG and the Animal Nutrition and Feeding Laboratory (referred to in Portuguese as LNAA), from the Department of Animal Science from the Federal University of Paraíba—UFPB—Campus of Areia, PB.

2.1 Protein Enrichment

Bioreactors were used in the semisolid fermentation of mesquite pods using the yeast *Saccharomyces cerevisiae* (commercial yeast, Fleischmann brand) with moisture of 70 %.

Table 1 Actual and coded levels of the input variables

Variables	Encoded levels		
	Level (−1)	Central point	Level (+1)
Yeast concentration (g L^{-1})	2	4	6
Temperature (°C)	30	35	40

Table 2 Matrix of 22 + 3 factorial design for semisolid fermentation

Tests	Yeast concentration (%)	Temperature (°C)
1	−1 (2 %)	−1 (30 °C)
2	+1 (6 %)	−1 (30 °C)
3	−1 (2 %)	+1 (40 °C)
4	+1 (6 %)	+1 (40 °C)
5	0 (4 %)	0 (35 °C)
6	0 (4 %)	0 (35 °C)
7	0 (4 %)	0 (35 °C)

The substrate (mesquite residue) was inoculated by yeast to obtain a semisolid fermentation in the bioreactors in concentrations of 2, 4 and 6 % and at temperatures of 30, 35 and 40 °C. At every given time, the samples were removed and stored in plastic containers with a capacity of 50 mL, which were conducted in oven at 55 °C for 24 h so as to avoid changes of nutrients, especially nitrogen compounds.

The actual and coded levels of independent variables, in order to study the influence of the input variables (the yeast concentration and temperature) are shown in Table 1. The 22 factorial design was employed with three trials at the central point (Table 2) of the planning matrix, having pH, protein and ART at temperatures 30, 35, 40 °C as response variables to moisture.

Figure 1 shows the process of semisolid state fermentation of mesquite bran in bioreactors in yeast concentrations of 2, 4, and 6 % and at temperatures of 30, 35 and 40 °C, according to the factorial design.

2.2 *Description of the Process of Semisolid Fermentation in Bioreactors to Obtain Protein Enrichment*

2.2.1 Solid Fraction (Bran)

Before being subjected to semisolid state fermentation, preliminary tests were performed to observe the growth of the microorganism to obtain the protein enrichment.

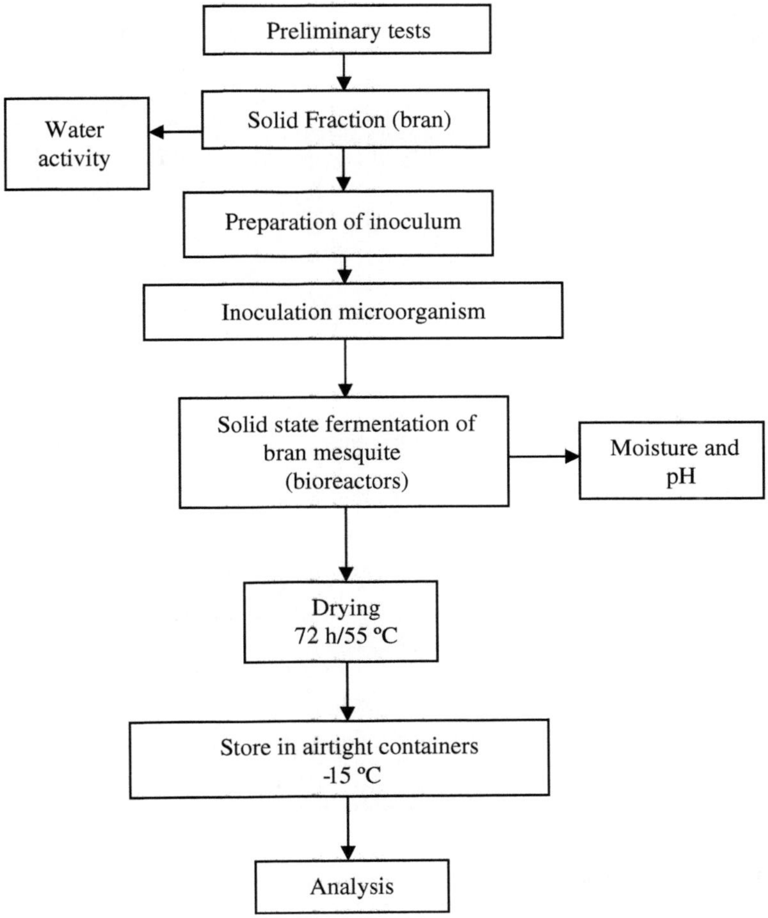

Fig. 1 Flowchart of experimental nutritional enrichment of mesquite residue

2.2.2 Drying and Milling

Drying of waste from mesquite pods was performed at 55 °C for 24 h in aluminum trays in the oven with air circulation. Then the material was ground in an electric mill (Tecnal brand, model TE.631) at the speed of 21,500 rpm to conduct grinding. The initial moisture content was around 8 % wb (Fig. 2).

2.2.3 Screening

Once crushed and milled, a homogeneous sample was obtained which was sieved onto mesh. The screening was performed using sieves of Tyler 60 and 80 (Fig. 3).

Fig. 2 Grinding of waste from mesquite pods

Fig. 3 Screening of waste from mesquite pods

Table 3 Size classification of flour and bran from mesquite pods

Classification	Thickness of sieve (mesh)	Thickness of sieve (mesh)
Bran	Above 60	Above 0.250

2.2.4 Grain-Size Measurement

In order to obtain the bran, a fraction was obtained with grain size (48 mesh). All material was sieved in the Granutest brand sieves with holes 0.25 (Table 3).

2.2.5 Moisture Adsorption Isotherms

For this achievement, the equipment Termoconstanter Novasina TH200 was used, at the temperature of 30 °C. Initially, 10 g of the waste was weighed (mesquite bran) in 10 containers and amount of distilled water with different amounts (0.2,

0.5, 0.8, 1.0, 1.5, 2.0, 2.5, 3.0, 3.5 and 4.0 mL). After the materials have been moistened and then homogenized, about 2 g of the sample were withdrawn and placed inside the cell for reading on the device. The reading of the water activity (a_w) was done in two stages: before and after equilibrium by Eq. 1.

$$X_{eq} = \frac{m_i - M_s}{m_s} \tag{1}$$

where:

X_{eq} Humidity (dry basis)
m_i Initial mass (g)
m_s Dry mass (g)

Once the data were found regarding water activity (a_w) and its balance moistures (X_{eq}), the adsorption isotherms was constructed for the temperature of 30 °C. GAB (Eq. 2) and BET (Eq. 3) mathematical models were used.

$$\text{GAB} = X_{eq} = \frac{X_m.C.K.a_w}{(1 - K.a_w).(1 - K.a + C.K.a_w)} \tag{2}$$

$$\text{BET} = \frac{X_{eq}}{X_w} = \frac{Ca_w}{1 - a_w}\left\{\frac{1 - (n+1)a_w^n + na_w^{n+1}}{1 - (1 - C)a_w - Ca_w^{n+1}}\right\} \tag{3}$$

where:

X_{eq} Equilibrium moisture content, g g^{-1}
X_m Moisture in the molecular monolayer
a_w Water activity
C, n, a, c Parameters of the models

Data adjustment was observed by means of the coefficient of determination (R2) and average percentage deviation (P) given by Eq. (4).

$$P = \frac{100}{n}\sum_{i=1}^{n}\frac{|VE - VT|}{VE} \tag{4}$$

where:

n Number of experimental data
EV Value of experimental moisture of the material
VT Theoretical value of moisture

2.2.6 Material Preparation and Inoculation on the Substrate

Rectangular trays of stainless steel (Fig. 4) with the dimensions of $35 \times 20 \times 45$ cm^3 were used to condition a volume of 3 L. These bioreactors

Fig. 4 Semi-solid fermentation in bioreactor trays

Fig. 5 Semi-solid fermentation of mesquite bran in the oven without movement

were taken to an oven without air circulation to begin the fermentation at temperatures of 30, 35 and 40 °C and the concentration of yeast (*Saccharomyces cerevisiae*) of 2, 4, and 6 %, according to the factorial design. This contained commercial yeast, with a humidity of 70 % on wet basis. For each tray 500 g of substrate were placed.

2.2.7 Semi-Solid Fermentation of Bran Mesquite

The analyses of pH and moisture for different times (0, 2, 4, 6, 12, 24, 48, 72, and 96 h) were performed in an oven without air circulation (Fig. 5). At each specified time the samples were placed in black plastic pots with a capacity of 50 mL samples which were conducted to the oven at 55 °C for 24 h in order to avoid changes in nutrients, especially nitrogen compounds.

2.2.8 Storage

All material was stored at 15 °C in film plastic pots so that afterwards the due physical and chemical analyzes were performed.

2.2.9 Physical and Chemical Analyzes

Moisture:

Moisture was determined following the methodology of Brazil [5].

pH:

In order to determine the pH of the material, 2.5 g of each sample (in triplicate) were used and these were placed in the Erlenmeyer flask with 25 ml of distilled water at an average temperature of 26 °C. The pH was determined according to AOAC [3].

Increased protein:

The crude protein content was determined by the Kjeldahl method using UV-Vis spectrophotometry. The protein content was obtained according to the nitrogen by multiplying it by a factor of 6.25 according to Eq. 5.

$$AP = \frac{(\%) Proteina\,Bruta(enriquecida) - Proteina\,Bruta(in.natura)}{(\%) Proteina\,Bruta(in.natura)} \times 100 \quad (5)$$

Total reducing sugars (TRS):

The percentage of total reducing sugars was determined by the DNS (3.5 Dinitrosalicylic Acid) method, according to AOAC [3].

Gross energy (GE):

The analysis of gross energy of the mesquite bran was performed according to AOAC [3].

Phosphorus (P), potassium (K), calcium (Ca) and magnesium (Mg):

The analyses of the percentage of P, K, Ca and Mg from the mesquite bran were determined as described by Silva [28] and also by Le Poidevin and Robinson [12].

Neutral detergent fiber (NDF) and acid detergent fiber (ADF):

The NDF and ADF analyzes were performed according to AOAC [3].

3 Results and Discussion

3.1 Physico-Chemical and Chemical Composition of the Bran in Natura from Mesquite Pods

The results of physico-chemical and chemical composition of the waste from mesquite pods in natura are described in Table 4.

Table 4 Physico-chemical composition of *a* bran in natura from mesquite pods

Component	Bran in natura
Moisture content (%)	7.87 ± 0.09
Total acidity (%)	6.75 ± 0.64
Protein (%)	7.65 ± 0.05
Lipíds (%)	5.49 ± 0.00
Minerals (%)	2.24 ± 0.01
NDF (%)	69.45
ADF (%)	51.98
Total reducing sugars (%)	87.75 ± 0.64
pH	4.73 ± 0.00
Organic matter	97.86
Dry matter (%)	92.13
Crude fat (%)	3.15
Ca (g kg^{-1})	2.69
Mg (g kg^{-1})	1.18
N (g kg^{-1})	20.30
P (g kg^{-1})	1.28
K (g kg^{-1})	3.64
Fe (g kg^{-1})	4.36

The moisture content for the residue of mesquite pods found in this study was lower than that described by Silva et al. [27], who found a mean value of 11.52 %, which is similar to the value found by Almeida et al. [1], i.e., 7.18 %. Knowledge of the bran moisture is crucial to efficiently and appropriately study protein enrichment, since by means of its control it is possible to avoid both contamination of the substrate due to the high moisture content or even a failure in growth of yeasts due to the low moisture content. Thus, protein enrichment requires balancing the moisture level of the material that will be used depending on the metabolism of the microorganism [20].

The total acidity of the waste from mesquite pods was similar to that found by Silva et al. [27], which was 6.84 %. However, the levels of lipids (1.54 %) and protein (6.84/6.17 %) disagree with those obtained in our work. Santos and Rebouças [24] found a value of 7.82 % for crude protein, agreeing with the result determined in our study. According to Bravo et al. [6], the minimum protein content required for feed for ruminants is 7.0 %. Therefore, in this regard, the product obtained in this work would agree to feed use for this type of animal, for example.

The increased intake of NDF and ADF can affect the digestibility of fibrous compounds by increasing the rate of passage of gut contents [14] and may result in a reduction in digestibility of NDF and ADF, when the replacement level of mesquite pod bran in concentrated feed is higher, because feed presenting high replacement levels also have higher levels of these compounds. The mineral content which was analyzed for the residue of mesquite pods was lower than those found by Silva et al. [27], i.e., 2.84 %, Santos and Rebouças [24], i.e., 3.82 %, and Silva [26] in the amount of 3.18 %.

The pH of the residue found in the sample of mesquite pods was 4.73. According to Teles et al. [29], the pH around 4.5 provides only the development of fungi and some yeasts, microorganisms used in this study. Also according to these authors, in relation to calcium and phosphorus micronutrients, the values were 0.78 and 0.27 respectively, lower than those determined in the present study.

The total reducing sugars are the soluble sugars present in the material, consisting of glucose, fructose and sucrose. The TRS content found in this study was 87.75 %, a value which is similar to Silva's [26], who studied the production of brandy mesquite, which was 88.63 %. This compound is of paramount importance for animal feed, being a carbohydrate source to be used as byproducts in fermentation processes.

3.2 Adsorption Isotherm of Bran Moisture (Residue) from Mesquite Pods in Natura

The experimentally obtained data, concerning the equilibrium moisture content (X_{eq}) at 30 °C, for the residue (Bran) of mesquite pods in natura and their respective values of water activity (a_w) are shown in Fig. 6.

Table 5 presents the parameters of the mathematical models GAB and BET, the coefficients of determination (R2) and the relative mean deviation (P) obtained for the adjustment to the experimental data of moisture adsorption isotherms of bran (residue) of mesquite pods in natura at the studied temperature.

From Table 5, it follows that all the models were appropriately adjusted to the experimental data, since the P value indicates a good fit when it is less than 10 %, while R2 should be as close to unity [13, 15].

According to the results, the values of the coefficient of determination (R2) and the mean relative error (P) from the GAB model best described the moisture adsorption isotherms of bran (residue) of mesquite pods in natura at temperature 30 °C, as compared to the BET model.

Ferreira and Pena [10] found a good adjustment to the data of the experiments using the GAB model for the adsorption isotherms of peach-palm flour at 15 and 35 °C. Almeida et al. [1] also found that the GAB model showed better adjustment to evaluate the adsorption isotherms of mesquite pods at the temperature of 30 °C with a coefficient of determination of 99.81 % and a standard deviation of 2.44 %.

Analyzing the data in Table 5, it can be seen that the GAB model showed value of the contents at the molecular monolayer (Xm) lower than the BET model, while for the heat of sorption (C) the value was higher. The constant (K) of the GAB model showed a value close to unit 1.

Figure 7 shows the moisture adsorption isotherm of waste from mesquite pods in natura regarding the temperature of 30 °C, adjusted by the GAB model. It can be seen that there was a satisfactory adjustment to the extent of the range of water activity (a_w).

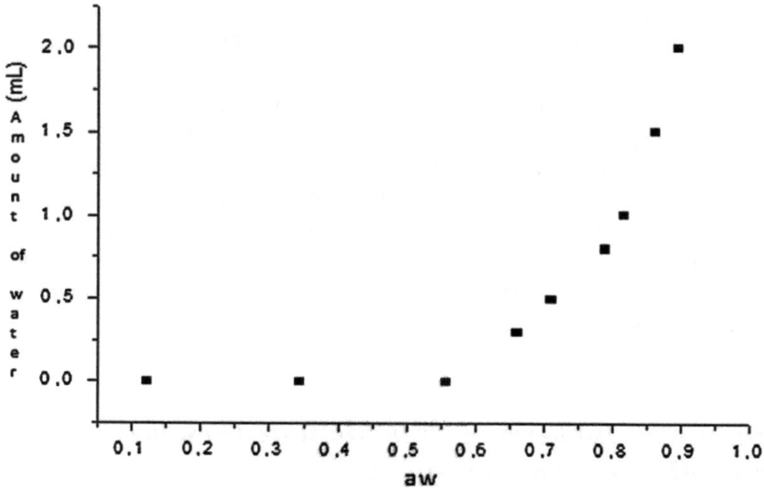

Fig. 6 Adsorption isotherm of moisture bran (residue) from mesquite pods in natura at temperature of 30 °C

Table 5 Tuning parameters of GAB and BET mathematical model, with the respective coefficient of determination (R2) and the mean percentage deviation (P), temperature 30 °C

Temperature (30 °C)	Parameters			R^2	P (%)
GAB	Xm	C	K		
	0.044	12.941	0.955	0.998	3.60
BET	Xm	C	n	0.979	4.28
	0.231	0.008	4.347		

In Fig. 8, it is possible to see the moisture adsorption isotherm of the residue of mesquite pods in natura related to the temperature of 30 °C, adjusted by the BET model. It was found that there was not a satisfactory adjustment of the length of the strip of water activity (a_w), only at the start and end points there were two points of water activity (a_w) that fit within the model.

It is observed in Figs. 7 and 8 that the equilibrium moisture content on a dry basis increases with the increasing of water activity. The optimum range of water activity for the growth of the yeast (*Saccharomyces cerevisiae*) which will be used in the enrichment process must be between 0.90 and 0.99 a_w of the substrate [9, 25]. Then, by correlating water activity and equilibrium moisture content shown in Fig. 8, it can be stated that the process of semisolid fermentation for the enrichment of mesquite pods should be operated with an initial moisture content (dry basis) greater than 0.30, which corresponds to a moisture content (wet basis) of more than 23 %, otherwise the growth of the microorganism is inhibited and may not even develop. For the storage of bran in natura, using neither additives nor cooling, it is necessary

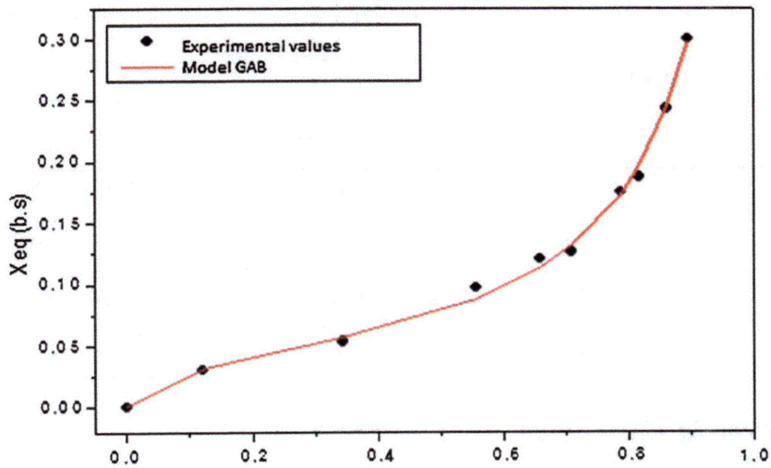

Fig. 7 Adsorption isotherm for the residue of mesquite pods in natura at the temperature of 30 °C (GA model adjustment)

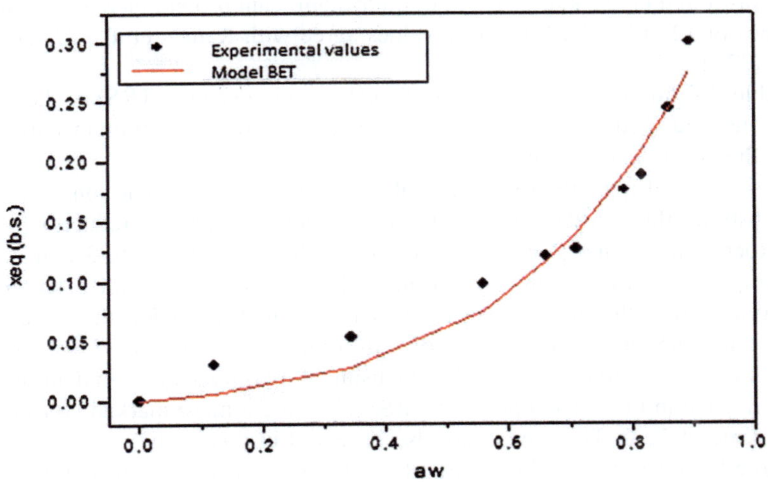

Fig. 8 Adsorption isotherm for the residue of mesquite pods in natura at the temperature of 30 °C (BET model adjustment)

for optimal activity range of residual water to be between 0.25 and 0.35, which corresponds to a moisture content (dry basis) below 0.053, equivalent to a moisture (wet basis) of 5.04 %, thus eliminating any growth of microorganisms [23].

Table 6 Values of crude protein during the process of semisolid fermentation with nine experiments

Experiment/Temp. (°C)	Fermentation time (h)								
	Protein content in natura (7.6)								
	0	2	4	6	12	24	48	72	96
1 (30)	9.66	11.08	9.46	10.47	15.33	10.07	10.47	11.89	11.28
2 (30)	11.82	13.51	14.93	16.14	17.15	13.10	12.09	12.29	12.09
3 (40)	10.88	16.14	10.07	16.14	17.76	23.64	16.55	16.95	16.14
4 (40)	13.71	20.40	20.19	21.41	17.15	26.27	22.22	23.64	22.22
5 (35)	12.50	13.51	12.90	13.71	10.47	13.10	11.69	14.12	10.88
6 (35)	12.29	13.01	11.48	12.09	12.29	12.90	12.09	14.32	14.12
7 (35)	11.89	12.09	11.48	11.69	8.65	10.07	9.05	15.74	13.10

3.3 Protein Enrichment

Table 6 depicts the fermentation kinetics of bran protein enrichment at 30 °C with concentrations of yeast of 2 and 6 %, respectively, where it can be observed that the best time the substrate was fermented by the yeast *Saccharomyces cerevisiae*.

The protein values found in this work are in accordance with the NRC [18].

It was found in this study (Table 6) quite significant protein increase, especially in times of 12, 48 and 72 h. These values agree with Ramos et al. [22] and Siva et al. [27] for feed.

Table 7 shows the data for levels of total reducing sugars (TRS) of the residue from mesquite pods throughout the fermentation process of protein enrichment using the yeast *Saccharomyces cerevisiae*.

It is observed from the results in Table 7 that there was sugar consumption of the substrates through the fermentation time. There is a gradual reduction of TRS for experiments 1 and 2 until the time of 96 h. In experiments 3 and 4, TRS consumption occurred more rapidly until the time of 24 h, with a softer consumption during the entire process. Almeida et al. [2] studied the behavior of consuming TRS during the process of semisolid fermentation at 38 °C, to investigate the bioconversion of mandacaru using Aspergillus niger and observed a slight decline in the consumption of TRS, following a more marked consumption at the time of 72 h. This trend was also observed in this study.

Table 8 shows the results of the pH in the process of protein enrichment of mesquite pod residue.

Analyzing the pH values (Table 8), it is observed that in experiments 1 and 2, the values in general increased until 48 h. In the experiments 3 and 4, there was also an oscillation of pH in the fermentation process over time, but at the time of 12 h, there was a greater increase in pH. For experiments 5, 6 and 7 at a temperature of 35 °C, greater pH increase at the time of 96 h.

Table 9 contains the moisture in humid basis in the process of protein enrichment of the residue of mesquite pods in different concentrations of yeast and temperature.

Table 7 Values of total reducing sugars during the process of semi-solid state fermentation

Experiment/Temp. (°C)	Fermentation time (h)								
	in natura (120.231)								
	0	2	4	6	12	24	48	72	96
1 (30)	105.87	84.3	60.40	37.66	36.64	32.24	31.22	26.81	6.45
2 (30)	110.80	52.26	32.58	13.24	11.54	11.88	7.13	6.11	4.00
3 (40)	111.80	70.77	68.54	38.00	22.73	12.89	10.52	10.86	9.84
4 (40)	114.56	74.00	89.58	32.57	20.69	13.23	9.50	9.50	8.48
5 (35)	118.09	81.94	60.40	20.36	20.02	19.00	12.21	10.85	9.16
6 (35)	118.00	84.90	52.93	19.68	12.63	16.29	13.24	11.87	11.19
7 (35)	117.67	81.03	50.56	32.24	20.69	20.36	19.00	17.64	10.18

Table 8 pH value in the process of protein enrichment of mesquite pod residue

Experiments/temp. (°C)	Fermentation time (h)								
	0	2	4	6	12	24	48	72	96
1 (30)	4.54	4.75	4.97	4.61	4.72	4.92	5.06	4.88	4.12
2 (30)	4.63	4.86	4.57	4.73	4.89	5.04	5.14	5.04	3.94
3 (40)	4.37	4.28	4.14	4.19	4.43	3.80	4.07	4.28	4.23
4 (40)	4.50	4.06	4.05	3.98	4.28	3.72	3.95	4.23	4.04
5 (35)	4.38	4.21	3.95	3.97	3.89	3.94	4.03	4.03	4.74
6 (35)	4.35	4.09	3.92	3.98	3.88	3.88	3.87	3.97	4.59
7 (35)	4.58	4.14	3.98	4.05	4.06	4.07	4.31	4.42	4.72

Table 9 Moisture content (w.b.) in the process of protein enrichment of waste from mesquite pod

Experiments/temp. (°C)	Fermentation time (h)								
	0	2	4	6	12	24	48	72	96
1 (30)	46.59	46.63	48.93	47.92	43.82	40.85	38.07	35.43	28.18
2 (30)	47.56	48.50	52.12	47.13	41.46	41.74	41.31	38.30	33.05
3 (40)	43.90	46.44	47.78	46.32	45.43	39.58	33.16	14.57	9.77
4 (40)	54.66	52.93	51.81	50.62	49.40	46.33	35.45	26.72	10.82
5 (35)	48.50	49.14	48.66	47.91	48.05	44.58	41.96	35.09	30.87
6 (35)	46.26	49.97	48.94	48.60	47.95	45.97	44.73	43.17	37.16
7 (35)	48.45	47.87	49.05	47.57	46.33	44.93	36.14	33.89	28.83

It is observed in Table 9 decreases of moisture in all fermentations, according to these figures presented. Small changes that occurred may have been due to the susceptibility of moisture exchange with the environment and air drying that occurs normally in biological products [1, 10].

4 Conclusions

1. The applied models fitted well to the experimental data, and the GAB model best represents the data.
2. The best protein value for residues of mesquite pods during semisolid fermentation using a commercial yeast *Saccharomyces cerevisiae* was in the time of 24 h at 40 °C with a yeast concentration of 6 % in the amount of 26 27 %, resulting in an increase of 345.65 % protein.
3. The values of total reducing sugars were mostly inversely proportional to the protein increase for all experiments.

References

1. de A. C. Almeida, F., da Silva, J.E., Araújo, M.E.R., de Gouveia, J.P.G., de Almeida, S.A.: Chemical components and study the moisture contente in mesquite pods. Braz. J. Agroind. Prod. **2**, 43–50 (2003) (In Portuguese)
2. Almeida, M.M., de S. Conrado, L., da Silva, F.L.H., Freire, R.M.M., Valença, A.R.: Characterization seafood mandacaru from two paraibanas cities. Braz. J. Agroind. Prod. **11**, 15–20 (2009) (In Portuguese)
3. AOAC—Association of Official Agricultural Chemists: Official Methods of Analysis, 12th edn. AOAC, Washington (1990)
4. Araújo, L.F., da SILVA, F.L.H., Oliveira, L.S.C., de Medeiros, A.N., Perazzo Neto, A.: Bioconversion of thornlees cactos tree (*Cereus jamacaru*) in alternative feed for ruminant animals. Agricultral Sci. Tecnol. **3**, 53–57 (2009) (In Portuguese)
5. Brasil. Ministry of Health. National Health Suverllaince Agency: Physicochemical Methods for Food Analysis, 1018p. Ministry of the Health, Brasília (2005) (Series A. Techinical Standarts) (In Portuguese)
6. Bravo, C.E.C., Carvalho, E.P., Schwan, R.F., Gómez, R.J.H.C., Pilon, L.: Determination of ideal Conditions for Poligalacturonase Production by *Kluyveromyces marxianus*. Sci. Agrotechnol. **24**, 137–152 (2000)
7. Campos, A.R.N., de Oliveira, M.M., Trindade, G.A., Dantas, J.P., de S. C. Oliveira, L., da Silva, F.L.: Scaling-up the processo f nutriente enrichment of cactus pear (*Opuntia ficus indica* Mill) by solid state fermentation. In XVI Brazilian Congress Chemical Engineering—BCCE. Lorena, 2006. 1608–1615
8. del Bianchi, V.L.D., Moraes, I.O., Capalbo, D.M.F.: Fermentation in semi-solid médium. In: Aquarone, E., Lima, U.A., Borzani, W. (eds.) Biotechnology: Biochemical Engineering, vol. 2, pp. 247–276 (2001)
9. Gould, G.W.: Drying, raised osmotic pressure and low water activity. In: Gould, G.V. (ed.) Mechanisms of Action of Food Preservation Procedures, 117p. Applied Science. Elsevier, London (1989)
10. Ferreira, C.D., Pena, R.S.: Hygroscopic behavior of peach palm flour (*Bactrisgasipaes*). Sci. Food Technol. **23**, 251–255 (2003)
11. Holanda, J.S., Oliveira, A.J., Ferreira, A.C.: Protein enrichment of cashew apple with the use of yeast for animal feed. Braz. Agric. Res. **33**, 787–792 (2001). (In Portuguese)
12. le Poidevin, L.A., Robinson, L.A.: Methods used in the diagnosis of leaf plantations Booken group in British Guiana: overall Sample and technical analysis. Fertileté **21**, 3–11 (1964)
13. Labuza, T.P., Kaanane, K., Chen, J.Y.: Effect of temperature on the moisture sorption isotherms and water activity shift of two dehydrated foods. J. Food Sci. **50**, 358–392 (1985)

14. LaCasha, P.A., Brady, H.A., Allen, V.G., Richardson, C.R., Pond, K.R.: Voluntary intake, digestibility, and subsequent selection of Matuabromegrass, coastal bermudagrass, and alfafa hays by yearling horses. J. Anim. Sci. **77**, 2766–2773 (1999)
15. Lomauro, C.J., Bakshi, A.S., Labuza, T.P.: Evaluation of food moisture sorption isotherm equations. Part I: fruit, vegetable and meat products. Lebensmittel–Wissenschaft Techonol. **18**, 111–117 (1985)
16. Muniz, M.B., Almeida, M.M., Sampaio, P.M., Rocha, A.S., da Silva, F.L.H.: Evaluation process of nutritional enrichment of mesquite pods (*Prosopolis juliflora* (SW DC). In: XVII. National Symposium Bioprocess, Natal (2009)
17. Muniz, M.B., Gomes, J.P., da Silva, F.L.H., Silva, C.G., Rocha, A.S.: Production of fermentet mesquite (*Prosopolis Juliflora*). In: Brasilian Congress of Chemical Engineering—COBEQ, 2012, Búzios, RJ (In Portuguese)
18. NRC—National Research Council: Nutrient Requirements of the Beef Cattle, 6th edn., vol. 1, p. 157. National Academy of Science, Washington (2000)
19. de Oliveira, M.M., Campos, A.R.N., Dantas, J.P., Gomes J.P., Silva, F.L.H.: Sorption isotherms of the agroindrustrial waste shell pineapple (*Ananas comosus* L. Mer). Braz. J. Agric. Environ. Eng. **9**, 565–569 (2005) (In Portuguese)
20. Perazzo Neto, A., de França, F.P., Barbosa, H.P.: Determination of parameters for protein enrichment of mesquite (*Prosopis juliflora*) with *Aspergillus niger*. Sci. J. Anim. Prod. **1**, 176–182 (1999) (In Portuguese)
21. Pinto, G.A.S., Brito, E.S., da Silva, F.L.H., Santos, S.F.M., Macedo, G.R.: Solids state fermentation: na alternative to the use and enhancement of agro-indrustial waste. J. Ind. Chem. **74**, 17–20 (2006) (In Portuguese)
22. de S. N. Ramos, L., Lopes, J.B., Figueiredo, A.V., de Freitas, A.C., Farias L.A., da S. Santos, L., Silva, H.O.: Cashew pulp in diets for broilers in the final phase: perfomance and carcass traits. J. Anim. Sci. **35**, 804–810 (2006) (In Portuguese)
23. Santin, A.P.: Study of Drying the Inactivation of Yeast (*Saccharomyces cerevisiae*), 150p. Master (Thesis). F Federal University of Santa Catarina, SC (2002)
24. Santos, C.L., Rebouças, G.M.N.: Ingestive behavior of Santa Inês sheep fed diets containing different levels of mesquite pod meal, vol. 14. In: Amazon Anual Meeting. SBPC, 2007, UFBA, Belém (In Portuguese)
25. Santos, S.F.M., Nóbrega, J.E., Pinto, G.A.S., Macedo, G.R., Silva, F.L.H.: Characterization of the dry residue of the cashew apple for their use as a substrat for solid state fermentation. In: National Symposium de Bioprocess, 15, Recife (2005) (In Portuguese)
26. Silva, C.G.: Optimization of the Production Process Brandy Mesquite and Utilization of Solid Waste in Food Products, 173p. Thesis (Ph.D), Federal University of Campina Grande, PB (2009)
27. da Silva, J.H.V., da Silva, E.L., Jordão Filho, J., Toledo, R.S., Albino, L.F.T., Ribeiro, M.L.G., Couto, H.P.: Energy values and effects of inclusion of integral flour of mesquite pods (*Prosopis juliflora*) in diets of laying hens. J. Anim. Sci. **31**, 2255–2264 (2002) (In Portuguese)
28. Silva, J.D.: Food Analysis (Chemical and Biological Methods), 73p. UFV, São Paulo (1998) (In Portuguese)
29. Teles, M.M.N., Neiva, J.N.M., Cavalcante, M.A.B., Cândido, M.J.D., Clemetino, R.H., Restle, J.: Fermentation characteristics and nutritive value of elephant grass silages containing by-products off annatto, cashew and mango. J. Anim. Sci. **39**, 1905–1910 (2010). (In Portuguese)

Evaluation of Cashew Apple Bagasse for Xylitol Production

F. C. S. Lima, F. L. H. Silva, J. P. Gomes, M. B. Muniz and A. M. Santiago

Abstract In this chapter is presented the general features and theoretical foundations about the topic, as its importance, consequence and transformation of the lignocellulosic residues in biotechnological processes; xylitol and its importance in biotechnological production, fermentation process and kinetic study. After, are presented the procedures of the experimental study that aimed to evaluate the potential of the liquor obtained from the cashew apple bagasse, for the production of xylitol, through the fermentation kinetics and fermentative parameter. The experimental work has shown that *C. guilliermondii* was able to consume xylose in all fermentations performed with greater production of xylitol of

F. C. S. Lima (✉)
Federal Institute of Education, Science and Technology of Pernambuco—IFET/PE, Campus Belo Jardim, Pernambuco, Brazil
e-mail: flavia.c.7@hotmail.com

F. L. H. Silva
Department of Chemical Engineering, Center of Technology, Federal University of Paraíba (UFPB), João Pessoa, Paraiba, Brazil
e-mail: flavioluizh@yahoo.com

J. P. Gomes
Department of Agricultural Engineering, Center of Science and Technology, Federal University of Campina Grande (UFCG), Campina Grande, Paraiba, Brazil
e-mail: josivanda@gmail.com

M. B. Muniz
Post-Graduate in Process Engineering, Center of Science and Technology, Federal University of Campina Grande (UFCG), Campina Grande, Paraiba, Brazil
e-mail: mbmmuniz@yahoo.com

A. M. Santiago
Department of Chemical, Center of Science and Technology, State University of Paraiba (UEPB), Campina Grande, Paraiba, Brazil
e-mail: angelasantiago@oi.com.br

J. M. P. Q. Delgado and A. G. Barbosa de Lima (eds.), *Transport Phenomena and Drying of Solids and Particulate Materials*, Advanced Structured Materials 48, DOI: 10.1007/978-3-319-04054-7_9,

2953.19 mg/L in 12 h of fermentation, obtaining volumetric productivity of xylitol, the factor and the efficiency of conversion of xylose to xylitol 0.108 g/L h, 90.70 and 98.99 %, respectively.

Keywords Xylitol · Acid hydrolysis · Fermentation

1 Introduction

The agroindustrial residues are situated among the greatest sources of biomass in the world. In Brazil, the amount of lignocellulosic residues generated annually is 597 million tons [1], generating considerable damage to the economic activities of the agroindustrial sector and to the environment.

The accumulation of these residues unleashes a succession of problems of environmental origin arising from the imbalance generated by the uncontrolled decomposition of this material, besides the loss of potential energy resources, generating an economic problem.

The transformation of agroindustrial residues by biotechnological processes allows the development of the science and technology in emerging countries where this type of material is abundant considering the fact these countries have large agricultural productions, raw materials and low production of manufactured materials.

The use of lignocellulosic residues in biotechnological processes requires some preparation steps by means of the acid or enzymatic hydrolysis to release the cellulose or hemicellulose fractions [2].

The prehydrolysis consists on a substantiated technique, quite widespread, of relatively low cost, which yields the release of pentoses (xylose and arabinose) and hexoses (mannose, glucose and others) constituents of hemicellulose. Once the xylose is the second most abundant sugar in the biosphere, hydrolysates obtained by this technique is characterized by becoming ideal for the production of several compounds, one of which is the xylitol. Furthermore, the biotechnological production of xylitol from xylose of the hemicellulose of agroindustrial residues not only presents aggregation of value to the residues but also a cheaper alternative for xylitol production since, by the chemical method usually used, its production has a high cost, by the number and type of purification stages of the xylose solution and separation of the catalyst employed [3].

Thus, the optimization of the alternative process of low cost for the biotechnological production of xylitol has become a global challenge and has been intensively investigated because it may contribute to the strengthening of various industrial segments such as pharmaceutical, odontological and food industries.

In this context, it is worth remembering that Brazil is known for its great potential of production of renewable resources such as agricultural residues (such as bagasse from sugar cane, rice straw, wheat straw, oat hulls and fruit peel), forest residues (such as wood shavings, sawmill dust and rest) and cultures such as willow and elephant grass [4]. Considering that these residues may contribute to the reduction of production costs, the cashew apple emerges as an alternative raw material for the production of xylitol because of its wide availability and high concentration of reducing sugars.

In the Brazilian Northeast the cashew, agribusiness produces annually about 200 tons of almonds and 2 million tons of cashew apple [5]. And, with the industrial use of the cashew apple of only 15 % of the total, large part is lost in the field. The use of cashew apple for xylitol production is justified by the wide availability in some states in the Northeast. Mainly in the state of Ceará, which accounts for 68 % of the national production of cashew, being a way to take advantage of the residue (bagasse) from the fruit avoiding the waste, which is around 85 %, making the cashew culture being more valued, generating employment and income.

2 Biomass

Currently, the energy needs of the planet are based mainly on fossil fuels (oil, natural gas and mineral coal). The reserves of these deposits are finite and their exploitation has provoked serious environmental problems [6].

The search for alternative renewable sources will intensify to supply the constant increase of energy demand and raw materials. One of the few sources that have the potential to meet these challenges of sustainability is the biomass, which appears to be a raw material with great potential, viable alternative against fossil fuels for the production of transportation fuels and chemical products, since it is the only material rich in carbon available on the planet beyond fossil [6].

To achieve these goals for sustainable development, biorefineries should exercise a dominant role in this millennium [7]. In this context, several researchers have demonstrated interest in developing alternative energy generation technologies that reduce dependence on fossil sources and are domestically available, renewable and less polluting [8].

The vegetable biomass and algae are very attractive candidates to play this role, as they are considered the only sustainable nutrient sources for the production of liquid fuels. As this is a renewable, abundant and inexpensive source that directly or indirectly, is produced by photosynthesis, fixing carbon from the atmosphere in resident carbon through the formation of carbohydrates present in the biomass composition.

2.1 Lignocellulosic Residues

The lignocellulosic biomass covers various agricultural residues (straw, peels, shavings, corn fiber and sugar cane bagasse), wood and residues from the paper

Table 1 Partial chemical composition of some lignocellulosic residues

Lignocellulosic residues	Cellulose (%)	Hemicellulose (%)	Lignin (%)	References
Corn leaves	37.6	34.5	12.6	[14]
Sugarcane bagasse	40.0	21.5	27.5	[15]
Marabou	39.5	21.7	32.1	[16]
Cashew apple bagasse	24.6	15.1	24.6	[17]

industry. The compositions of these materials vary and are constituted mainly of cellulose (35–50 %) followed by hemicellulose (20–35 %) and lignin (10–25 %) and can be converted into energy, chemical products and food [9]. For this to occur it will be necessary, hydrolyze them, because due to its polysaccharide nature, are not directly utilized by microorganisms of industrial interest.

From technological perspective the sugars contained in cellulose fractions (glucose) and hemicellulose (xylose, arabinose, glucose, mannose and galactose) represent the substrates that can be used for the production of xylitol by fermentation pathway. However, the association between the three main fractions (cellulose, hemicellulose and lignin) is so great, imposing obstacles for the recovery of constituent sugars in the form of monomers with high purity grade [10].

The constituent cellulose that is the most abundant of the vegetal cell wall is a linear homopolymer of high molecular weight which contains, as a fundamental unit exclusively the β-D-glucose united among them, forming the cellobiose, compounds of repeating units in its chain. It presents as microcrystalline structure, which hardly dissolves or hydrolyzes under natural conditions [11].

The hemicelluloses are polymers whose composition may appear in condensed varied proportions, the backbone of their chains may be a homopolymer (usually consisting for a single repeating unit of sugar) or an heteropolymer (mixture of different sugars), as other components, such as acetic, glucuronic and ferulic acid. The hemicelluloses have low molecular mass and contain no crystalline regions and are therefore more susceptible to chemical hydrolysis under milder conditions and are closely associated to cellulose in the plant tissues [11, 12].

While lignin is formed from three main precursors: the *p*-coumaryl, coniferyl and sinapyl alcohols. Found in the cell wall of all vascular plants, is a phenolic polymer, derived from aromatic alcohols that have the function to confer rigidity and impermeability to water, besides mechanical and microbiological resistance to vegetal tissues [13]. It is presented in Table 1, the chemical composition of some lignocellulosic residues.

Many lignocellulosic materials such as organic residues [18–20], are used as a mean of cultivation to the biotechnological obtainment of xylitol. Among the different biomasses that compose the lignocellulosic materials, the cashew apple bagasses are promising sources for its bioconversion, since they constitute in abundant residue.

Fig. 1 Representation of the cashew parts

2.2 The Cashew Apple

The lignocellulosic biomasses represent an abundant source of sugars through biotechnological processes, can be converted into products of industrial interest given to the challenges of sustainability, context in which the culture of cashew tree has great potential for the technological development from their industrial residues known popularly as cashew bagasse that, in general, are reused to enrich animal feed or discarded for lack of incentive of its use as human food [21].

According to [22] the cashew tree (*Anacardium occidentale* L.) occupies a prominent place among the tropical fruit plants in face of the increasing commercialization of its main products. Belonging to the *Anacardiacea* family is, in turn, a typically Brazilian plant as it was already here, in the Brazilian Northeast, when the first Portuguese settlers arrived and from here spreaded their seeds for their domains, in Africa and Asia. The name cashew is derived from the indigenous word "acaiu", in Tupi, means nut which produces itself [23].

Cashew is a pseudo fruit often confused with the fruit, chestnut, a drupe whose seed is edible. Scientifically termed floral peduncle, with variant color between yellow and red, the peduncle has developed in a different way, is consumed in nature and appreciated for its juiciness [24]. In the innermost part of the nut is located the almond that is constituted of two fleshy and oily cotyledons, which compose the edible part of the fruit (Fig. 1).

Due to its fragility the cashew apple, is highly perishable when stored at room temperature and after a period of 48 h it presents wrinkled, ferments and consequently loses the attractiveness providing mechanisms to accelerate the microbiological degradation, thus contributing to the rejection or loss of hundreds of thousands of tons per year of the product [25].

2.3 Acid Prehydrolysis of Lignocellulosic Materials

The lignocellulosic materials may be treated by biological, physical, chemical or enzymatic processes, so that the release of fermentable sugars occurs. Among these, acid hydrolysis is the oldest and most used for the production of xylitol [26].

Diluted acids may be used for partial hydrolysis of lignocellulosic materials, a procedure called prehydrolysis or pre-treatment which consists in fractionating the chemical constituents most susceptible to the acid treatment, in order to remove the hemicellulose fraction (xylose, arabinose and other sugars) favoring the opening of the physical structure from the vegetal cell wall [26–29].

According to [10] the process of prehydrolysis with the use of dilute acids presents two categories: The pre-treatment at high temperatures (greater than 160 °C) continuous process and low solid loading (5–10 % mass of substrate/mass of the mixture) and low temperature (lower than 160 °C) batch process and high solid loading (10–40 %). These treatments allow achieving high yields and that at high temperatures, there is favoring of the hydrolysis of cellulose, while under milder conditions occurs higher conversion of xylans to xylose.

During the pretreatment with diluted acids such as sulfuric, nitric, acetic and hydrochloric acid, used as catalysts has been cited as the best type of pre-treatment for agroindustrial residues [30].

2.4 Toxic Effect and Treatment for the Removal of Inhibitors from the Fermentation Process

The biotechnological obtaintion of the xylitol presents restrictions as to the efficacy during the fermentation process, due to the presence of inhibitory compounds to microbial metabolism causing low yield and productivity in ethanol and xylitol, being necessary detoxification treatment before fermentation, to improve the performance [27].

The concentrations of these inhibitors (furfural, HMF and acetic acid) present in the liquor after the pretreatment, deserve attention, because they depend on the nature of the lignocellulosic material used and may limit the consumption of the carbon source, reducing the speed of growth (inhibiting agent) or even prevent bioprocess (toxic agent). In this context, the detoxification intended remove toxic compounds [31].

The detoxification by adsorption with activated charcoal is widely used for simple or combined treatment with alkalinization. Activated charcoal is added to the hydrolysate and kept under stirring for determined time and temperature. In this process, the adsorption capacity of compounds varies with the adsorption conditions and with the type of active charcoal employed. The main disadvantage of this method is the limited reuse of charcoal since its adsorption capacity is exhausted in an irreversibly way, in a few cycles of operation [30].

A new method currently employed for detoxification of lignocellulose hydrolysates, is to use the residual lignin of the hydrolysis as an adsorbent in the extraction, taking advantage of their hydrophobic properties. The advantage of using the residual lignin as detoxification agent front of chromatographic resins is mainly economic. Because the lignin is a byproduct of hydrolysis after having been used for detoxification still can be used as a primary fuel [32]. After the detoxification of the hydrolyzed liquors, they can be used in the bioconversion process of xylose to xylitol efficiently.

3 Xylitol

Industries that produce sweeteners have registered an increasing demand by the consumers for products with low sugar content and calorific value, among them, xylitol as a sugar substitute comes with some interesting physical and chemical properties that make it an important compound of high value to the industries: pharmaceutical, odontological and food [33].

Xylitol is a five carbon polyalcohol, which presents the molecular formula $C_5H_{12}O_5$ (1,2,3,4,5 pentahydroxy pentane) and molecular weight of 152.15 g [34]. The name xylitol is related to the xylose, sugar of the wood, from which xylitol was obtained by the first time. With open structure, the xylitol molecule has five hydroxyl groups (OH), each attached to a carbon atom which is why this compound is known as acyclic polyhydroxy alcohol or pentitol [35].

Xylitol presents properties similar to sucrose [36] besides being considered an atoxic substance, according to the Food and Drug Administration (FDA) regulating organ of food in the United States, as additive of the type GRAS (Generally Regarded as Safe), its incorporation in food is legally allowed.

Studies have reported that xylitol may have numerous medical applications however, [37] reported its use new applications in the medical area as agent for the prevention of acute otitis media in children. Have [38] developed a dental veneer containing 10 % xylitol, which has shown effective against caries, within 72 h after application and from there came a new product against caries and infections of the upper airways.

In the food industry the xylitol can be applied isolated or in association with other sweeteners in canned products, baked goods, jams, marmalades, jellies, desserts, chewing gum, ice cream and soft drinks, among others, improving color, texture, flavor and the stability of these products [39]. The high heat of negative dissolution responsible for the pleasant sensation of freshness and other physicochemical properties of xylitol, are presented in Table 2.

3.1 Production by Means of Biotechnological

According to Cassales et al. [40], an alternative to the formation of products with high added value through lignocellulosic hydrolysates, within the concept of biorefinery, is the production of xylitol. This pentahydroxylated alcohol has great economic importance because it can be used as a substitute for conventional sugars.

Table 2 Characteristics and physicochemical properties of xylitol

Properties	Characteristics or values
Empirical formula	$C_5H_{12}O_5$
Molar mass	152.15 g/mol
Appearance	Crystalline powder
Color	White
Flavor	Sweet
Odor	None
Melting point	92–96 °C
Boiling point	216 °C (1 atm)
pH (aqueous solution at 10 %)	5–7
Density (aqueous solution)	10 %—1.03 g/mL; 60 %—1.23 g/mL
Solubility in water at 20 °C	63 g/100 g solution
Viscosity (aqueous solution at 20 °C)	10 %—1.23 cP; 60 %—20.63 cP
Dissolution heat (endothermic)	−34.8 cal/g
Caloric value	2.4 kcal/g
Refractive index (25 °C)	1.3471 (aqueous solution at 10 %)
Hygroscopicity	At higher RH is more hygroscopic than the sucrose and less than the sorbitol
Sweetness	Similar to sucrose, superior to sorbitol and mannitol
Stability	Stable at 120 °C (not caramelizes)

Source [59]

The employment of hemicellulose hydrolysates with culture medium for the biotechnological development process of xylitol has been shown to be possible. The process is easier than the chemical conversion of xylose where the xylitol is traditionally produced, with low yield (50–60 %) and high cost [38]. However, the biotechnological production presents advantages such as use of milder conditions of temperature and pressure [41] not being necessary the use of pure xylose, since the microorganisms are capable of converting xylose to xylitol directly from the hemicellulosic hydrolysate [42].

As an alternative to conventional acquisition process, xylitol can be produced biotechnologically using yeast and/or enzymes. This process consists in the fermentation of hemicellulose hydrolysates (solutions rich in xylose) obtained from agroindustrial residues without the need for prior purification of the substrate and may compete with the traditional chemical process [43].

Among microorganisms, the yeasts *Candida guilliermondii* and *C. tropicalis* are considered the best producers of xylitol. Once inside the cell of the yeast, the xylose is reduced to xylitol by the enzyme xylose reductase (XR) dependent or NADH or NADPH. The xylitol formed relatively stable can then, be excreted from the cell or oxidized to xylulose by the dehydrogenase xylitol enzyme with participation of NAD^+ or $NADP^+$. The produced xylulose is phosphorylated by kinase xylulose enzyme, forming xylulose-5-phosphate, which is metabolized in the phosphopentose pathway [44].

Various other commercial products are obtained by fermentation highlighting the ethanol used, even as a fuel. Among the advantages presented by biotechnological processes, the cost reduction in relation to chemical processes stands out, besides presenting characteristics that minimize environmental impact, as decrease of toxicity of effluents and the use of renewable resources [43].

3.2 Fermentation Process

The fermentation can be defined as an alternative anaerobic process used by certain species of microorganisms. In this process, substrates can be fermented into products of commercial interest by the action of viable microorganisms. The sugars are excellent fermentable substrates because produce intermediates which can be reduced [45].

In general, the fermentation processes are classified according to the submerged fermentation or solid state. Crucially, one of the most exalted parameters on the difference of these two types of processes is the content of water present in the reaction medium [45]. According to these authors, in the solid state fermentation (SSF) there is an absence or near absence of free water. The present water in these systems is complexed with the solid matrix of the substrate or as a thin layer absorbed by the surface of the particles.

In submerged fermentation (SF) using a liquid fermentation medium in which the nutrient sources used are soluble, being considered a technique that has relative ease of large scale cultivation, because it ensures the homogeneity of the medium and ease control of the parameters of process, especially if monitored by appropriate sensors [46].

For that in the submerged cultivation occurs good cell proliferation, the liquid substrate must contain a source of carbon (energy source) as example: synthetic sources (glucose, xylose, maltose, lactose, sucrose, oat xylan, citrus pectin and glycerol) or natural sources (sugarcane bagasse, orange bagasse, oat bran, wheat bran) among others [46]. And a source of nitrogen because in a culture medium there must be some necessary vitamins for the metabolism of the microorganism such as yeast extract, the tryptone and the protease peptone.

In the hydrolysed means of the vegetal biomass the fermentation can be conducted in batch, feed-batch or continuous. The appropriate choice will depend on the kinetic properties of the fermentation agent on the substrate and the economic viability of the desired product to be obtained [47].

According to Prakasham et al. [48], the process of producing xylitol bioconversion in laboratory scale is often performed in batch using shaken flasks or reactor of the type stirred tank, with free or immobilized cells, and mediums based on xylose or hemicellulose hydrolysates, resulting in productivity values ranging between 0.55 and 0.78 g/L h.

Table 3 Kinetic parameters of fermentation

Term	Equation
Monod's equation	$\mu_x = \frac{\mu_{max} S}{K_s + S}$
Conversion efficiency	$\eta = \frac{Y_{P/S\ obtained}}{Y_{P/S\ theoretical}} \times 100$
Volumetric productivity	$Q_P = \frac{\Delta P}{\Delta t} = \frac{P_f - P_i}{t_f - t_i}$
Xylose consumption	$Y(\%) = \frac{S_i - S_f}{S_i} \times 100$
Conversion factor of xylose to xylitol	$Y_{P/S} = \frac{\Delta P}{\Delta S} = \frac{P_f - P_i}{S_i - S_f}$

μ is the rate of cell growth, μ_{max} is the maximum rate of cell growth, S the concentration of the limiting substrate and K_S the Monod's constant

3.3 Studies of the Fermentation and Kinetic Parameters

The study of the kinetics of microbial processes is subject of interest of research centers, because it aims to quantify the rate of cell growth, substrate consumption, product formation and other related parameters, besides to evaluate the influence of external factors such as pH, temperature, inhibitors etc. [49].

The mathematical models developed to represent the kinetics which occur in the fermentation process are based specifically on knowledge of the biochemical pathways that lead to the production of these compounds and the type of fermentation [50]. These models are essential for the development of strategies for optimization and control of fermentation processes.

The most simple and known equation to describe microbial growth is the Jacques Monod's equation, which considers the presence of the limiting substrate for the growth, described in Table 3 [51].

4 Obtainment of Xylitol

The present research was conducted at the Laboratory of Heat and Mass Transfer in Porous Media and Particulated Systems (LTCMMP) and at the Laboratory of Biochemical Engineering (LEB) of the Academic Unit of Chemical Engineering UFCG.

4.1 Raw Material

The raw materials used in this work were the cashew apple bagasse (*Anacardium occidentale* L.) and its prehydrolysed liquor.

4.1.1 Obtainment of the Cashew Apple Bagasse

The cashew apple bagasse was purchased in the juice production industry, Frutnat located in Campina Grande, Paraíba. This raw material was transported to the LTCMMP, where the following treatment steps occur: two washes of the cashew apple bagasse with distilled water at a temperature around 50 °C, for the average time of 20 min each wash, then, the bagasse was subjected to two additional washes with distilled water at room temperature for the leaching of remaining sugars during juice extraction, until it reaches zero °Brix.

After these steps the bagasse was placed on aluminum trays and taken to dry in an stove with air circulation at a temperature of 55 °C for 48 h. Then the material was removed, ground in a cutting mill and sieved at 48 mesh, to reduce the granulometry, and then vacuum packed in polypropylene plastic bags. They were then stored in hermetically sealed containers for use in research.

4.1.2 Obtainment of the Prehydrolyzate Liquor

The prehydrolysed liquor to the research in development was obtained from the raw material "cashew apple bagasse". This cashew residue was subjected to the prehydrolysis process in which the main fraction that was removed was the hemicellulosic, in a pressurized stainless steel reactor with FE50RP thermal controller (automated time system, heating and indoor/outdoor temperature) and 700 mL of capacity. The reactor was charged with the bagasse using a mass ratio of 1:6 and the sulfuric acid solution diluted at 3 % v/v, 95 % purity at 105 °C for 1 h, according to the study [52]. The liquor resulting from this prehydrolysis was obtained after filtration to the separation of the solid fraction constituted of cellulose and lignin.

4.2 Concentration of the Prehydrolysed Liquor

The prehydrolysed liquor was subjected to the process of concentration in a rotary evaporator connected to a vacuum pump. The working temperature of the rotary evaporator was 70 ± 5 °C in order to increase the sugar content, mainly xylose. The prehydrolysed was concentrated by reducing its initial volume (FC 2.5) and stored in a cold chamber at 4 °C for later use.

4.3 Detoxification Process of the Prehydrolysed Liquor

The liquor obtained from acid prehydrolysis, had its initial pH (0.98) elevated to the final pH of 5.0, with addition of NaOH 17.5 %; the liquor was then subjected to removal process (detoxification) employing up as adsorbent residual lignin.

The operating conditions were: mass ratio of 1:100 (1 g of adsorbent/100 g prehydrolysed liquor) keeping the mixture under stirring at 150 rpm in Shaker equipment at 30 °C for 1 h and vacuum filtered to the elimination of precipitate (Fig. 4) according to [53]. After each test the samples were collected for analysis of sugars and fermentation inhibitors (congeners) in (HPLC).

4.4 Study of Fermentation Process

4.4.1 Obtainment of Microorganisms

Initially, the microorganism was cultured on corn for obtaining a large amount of spores. Then the spores were dispersed. The yeast used in research for xylitol production was *CCT 3544 Candida guilliermondii* obtained in André Tosselo Foundation—FAT Collection of Tropical Cultures.

4.4.2 Reactivation of Lyophilized Cells

For reactivation of the cells was added 0.2 ml of distilled water with the aid of a Pasteur pipette, in the ampoule with the lyophilized yeast, in a sterile environment, then was made a suspension of cells, leaving them rehydrate for about 10–15 min, period after which the entire content was transferred of the ampoule to a test tube containing 5.0 ml of liquid culture medium YMA—"Yeast Malt-Extract" and incubated at 28 °C.

4.4.3 Growth of Cells on Solid Medium

From the culture grown in the broth was carried out a pealing on Petri dish containing the solid medium "Yeast-Malt Extract Agar" and incubated at 28 °C for 48 h, after the growing the Petri dishes containing yeasts were stored in refrigeration (2–8 °C).

4.4.4 Inoculum Preparation

Pealed cells of the CCT 3544 *Candida guilliermondii* on maintenance medium were transferred under aseptic conditions, with the aid of a platinum loop to test tubes containing about 5 ml of sterile distilled water. 1 mL aliquots of this suspension were then transferred to 125 mL Erlemneyer flasks containing 50 mL of semisynthetic culture medium, consisting of: (30 g/L of xylose, 2 g/L of ammonium sulphate, 0.1 g/L calcium chloride and 20 g/L extract solution of rice bran).

Table 4 Levels in real and coded values for the investigated factors in the 2^4 factorial planning with three replications at the center point

Variables (g/L)	Levels		
	−1	0	+1
Ammonium sulphate (g/L)	1.0	2.0	3.0
Calcium chloride (g/L)	0.5	1.0	2.0
Rice bran (g/L)	5.0	12.5	20.0
pH	4	5	6

All solutions were prepared separately; solutions of xylose and extract of rice bran were autoclaved at 111 °C for 15 min, whereas the stock solutions of $(NH_4)_2SO_4$ and $(CaCl_2)2H_2O$ at 121 °C, for 20 min.

Then, the Erlenmeyer flasks were incubated in a rotary agitator of the "shaker" type, under the following conditions: 200 rpm stirring at 28 °C for 24 h and pH 5.5, after this period the cells were recovered or separated by centrifugation at 2,000 xg, washed with sterile distilled water, centrifuged again, and after discarding of the supernatant were used to prepare a cell suspension of approximately 50 g/L.

4.4.5 Preparation of Cell Suspension

After washing with sterilized water and the centrifugation process, the supernatant was discarded, and then were weighed 50 g and the mass diluted in 1 L of water, and from this suspension was calculated the required volume to inoculate the medium with an initial concentration of 3 g/L in the medium or 107 cells/mL in all fermentations.

4.4.6 Fermentation Trials

The fermentation trials were conducted in 125 mL Erlenmeyer flasks containing 50 mL of fermentation medium, 3 g/L of inoculum and supplemented according to the experimental delineation presented in Table 4. The flasks were kept in an incubator of rotational movement under agitation of 200 rpm and temperature of 28 °C, by the time of 0–96 h performing the fermentation kinetics.

In Table 4 are the levels of the factors used in the planning in which (−1) and (+1) denote the lowest and highest levels, respectively, and (0) means the level of the central point.

4.5 Results and Discussion

4.5.1 The Kinetics of Fermentation for the Production of Xylitol

In Figs. 2, 3, 4, 5, 6 and 7 it is observed the kinetics of xylose consumption and the obtainment of xylitol, by means of the *Candida guilliermondii* yeast under the conditions of the 2^4 factorial planning with three replications at the central point in function of time.

The *C. guilliermondii* yeast was able to metabolize xylose in all fermentations performed. It has been observed in Figs. 2, 4 and 6, in which xylose was completely consumed between 72 and 96 h in the experiments (1, 7, 13 and 15). Studying the production of xylitol in hemicellulosic hydrolysates by [54] has observed a total consumption of xylose in 72 h of fermentation, both for the supplemented hydrolysate as for the synthetic medium.

According to Figs. 2, 4, 5, 6 and 7, between 36 and 48 h of fermentation the xylose consumption was slowly reduced by the yeast experiments (3, 8, 10, 12, 13 and 16 reaching an average of approximately 90 % probably due to the beginning of the exhaustion of xylose). Bier et al. [55] observed, by studying the growth and xylose consumption by *Candida guilliermondii,* a consumption of 90 % of xylose at 48 h of fermentation.

It is also verified that the smaller value of the pentose assimilation was obtained in experiments 3, 4, 7, 8 and 14 for the first 12 h of fermentation with xylose consumption less than 35 % (Figs. 3, 4 and 6), due to the presence of glucose inhibiting xylose metabolism. Fonseca and Faria [54] observed that the xylose consumption speed in synthetic medium, which contained no D-glucose, was higher than in hydrolysate and this fact justified by the presence of the hexoses in the fermentation medium which inhibited the xylose metabolism by repression, inactivating the transport system of xylose.

As for the xylitol obtainment (Figs. 2 and 8) it is noted that, initially, and for all the studied experiments the xylitol formation was less than 3000.0 mg/L, the maximum xylitol production was achieved in experiments 1, 2, 5 and 6 with concentrations from 2403.51 to 2953.19 mg/L, and consumption from 98.03 to 99.91 % of the sugars present in the medium (Figs. 2 and 3). This difference can be attributed to supplementation with rice bran extract, because it is a source of nitrogen which allows improving the metabolism of the yeast. Moutta et al. [56] observed this difference and attributed to the supplementation with yeast extract when studying the bioethanol production in hydrolysates of sugarcane bagasse.

Fonseca and Faria [54] found that the highest values of xylitol concentration were 40.5 g/L for synthetic medium and 37.7 g/L for the hydrolysate, after 72 h of fermentation, unlike this work where the decline of xylitol concentration was observed between the times of 48–72 h.

In Figs. 2, 4, 5 and 6 after 72 h of fermentation of the experiments (3, 9, 11 and 15) and 48 h of the experiment 13, the xylitol concentration in the fermentation medium gradually decreased. This reduction in the xylitol concentration can be attributed to a

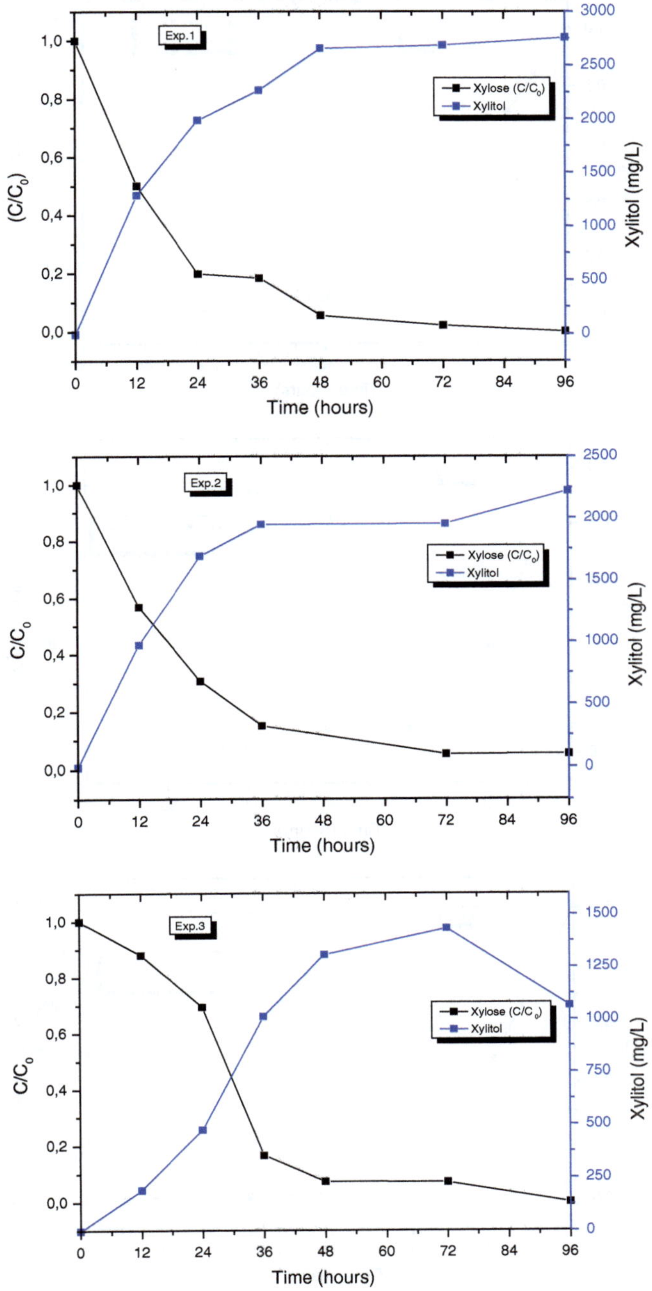

Fig. 2 Kinetic profiles of fermentations of the xylose consumption and xylitol production by using the yeast CCT 3544 *C. guilliermondii*, exp. 1–3

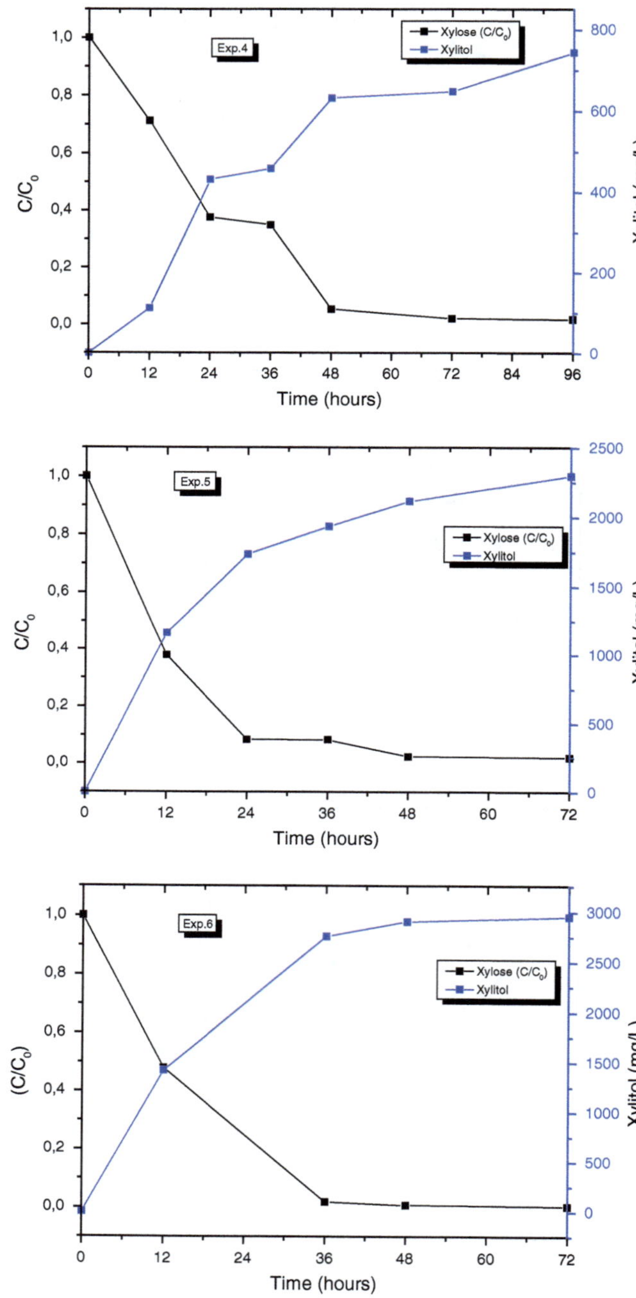

Fig. 3 Kinetic profiles of fermentations of the xylose consumption and xylitol production by using the yeast CCT 3544 *C. guilliermondii*, exp. 4–6

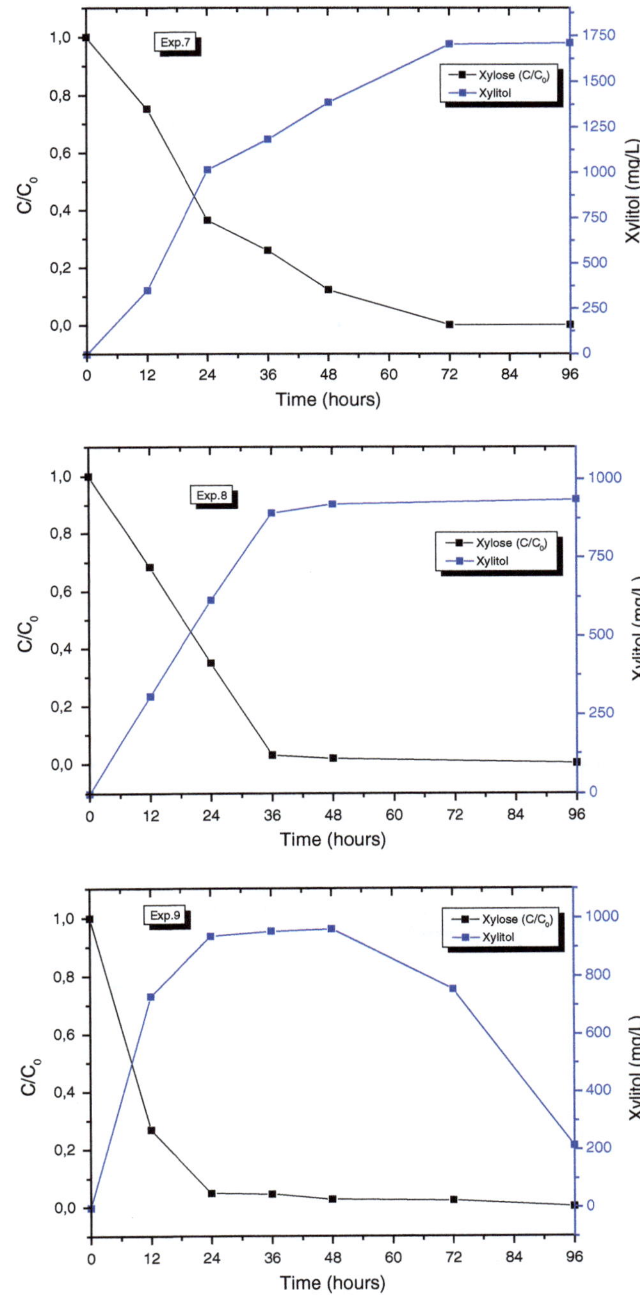

Fig. 4 Kinetic profiles of fermentations of the xylose consumption and xylitol production by using the yeast CCT 3544 *C. guilliermondii*, exp. 7–9

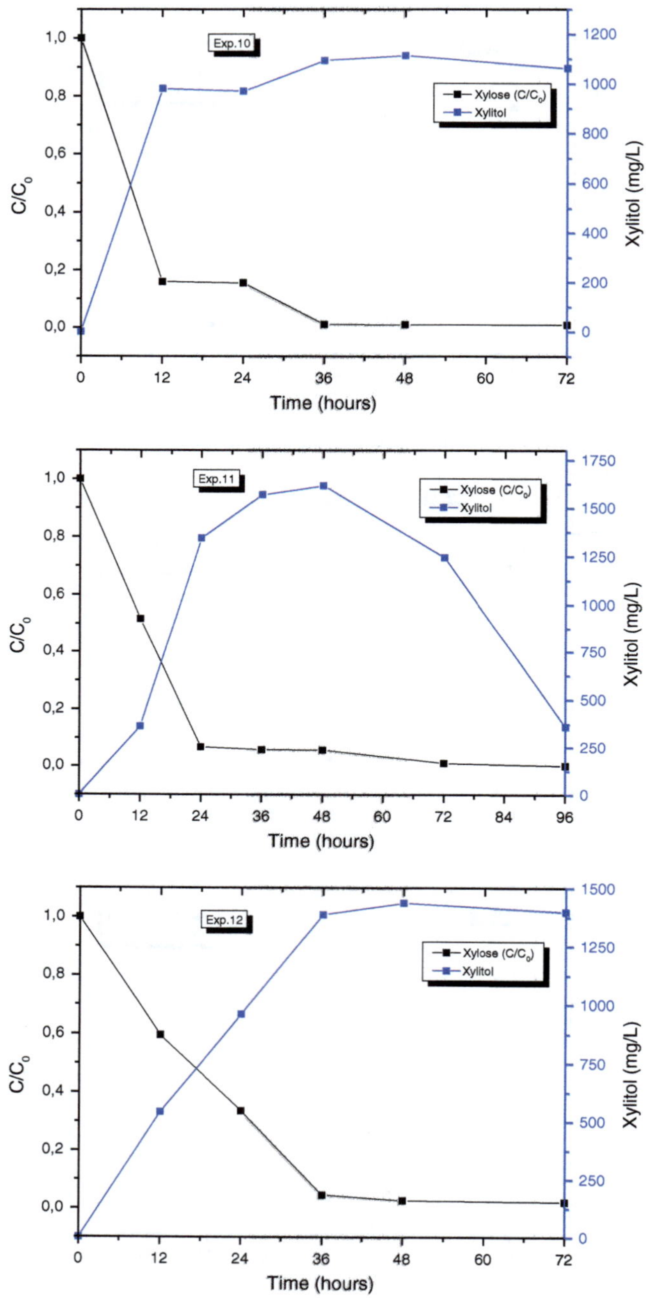

Fig. 5 Kinetic profiles of fermentations of the xylose consumption and xylitol production by using the yeast CCT 3544 *C. guilliermondii*, exp. 10–12

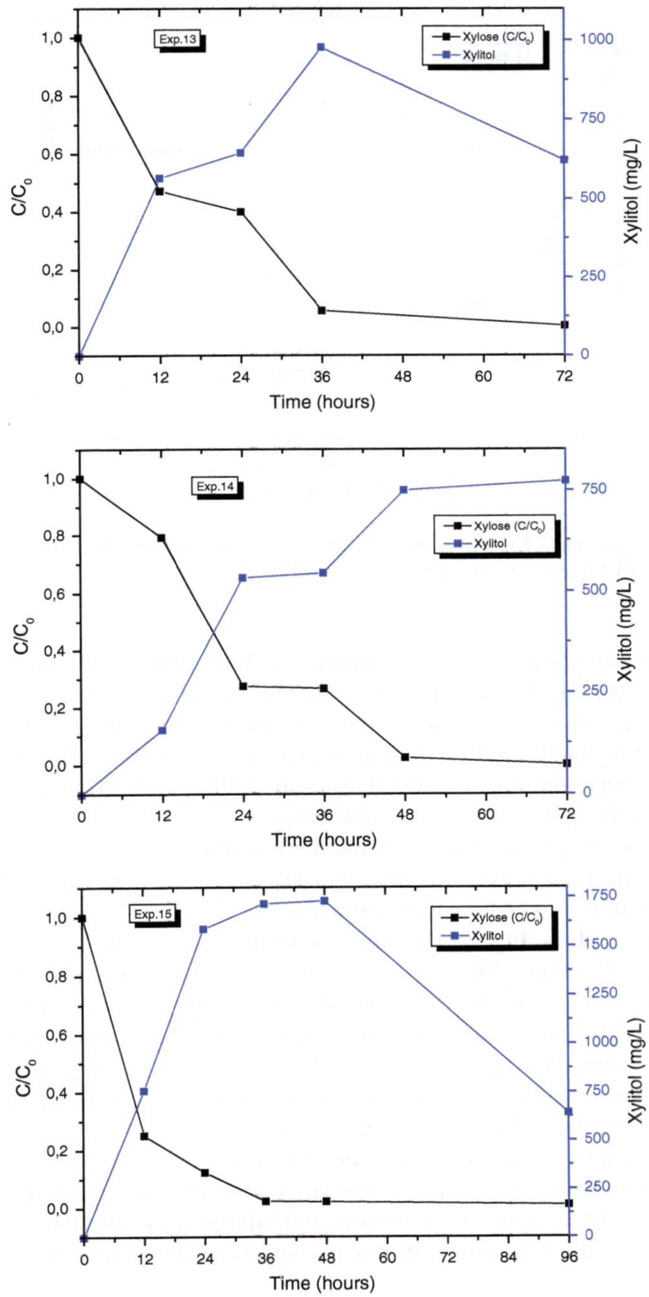

Fig. 6 Kinetic profiles of fermentations of the xylose consumption and xylitol production by using the yeast CCT 3544 *C. guilliermondii*, exp. 13–15

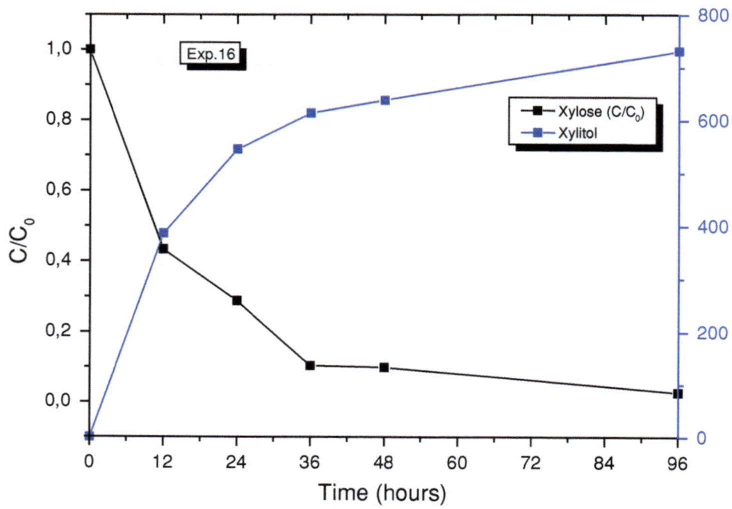

Fig. 7 Kinetic profiles of fermentations of the xylose consumption and xylitol production by using the yeast CCT 3544 *C. guilliermondii*, exp. 16

probable consumption by the yeast, due to the lack of sugars in the fermentation medium. Silva et al. [18] found, by researching the biotechnological production of xylitol in sugarcane hydrolysates, a decrease in the concentration of xylitol after 72 h, indicating that the xylitol polyol may be assimilated by the *C. guilliermondii*, thereby the reduction of the yield values of the xylitol production may be directly associated to the xylitol consumption by the yeast.

Bier et al. [55] discarded the fermentation times over 72 h for the xylitol production, since it observed that after this period the yeasts tended to consume the produced xylitol, thus reducing the biomass.

It is observed in Fig. 8, the kinetics of the fermentation process for the experiments 17, 18 and 19 corresponding to the central point, performed in triplicates in order to verify the reproducibility of the data. It was verified a similar behavior as to the xylose consumption to the experiments 17 and 18 that after 36 h showed a consumption of approximately 97 %, in other words practically all the xylose was consumed, however, to the experiment 19 this consumption was slower 95 % due probably to the xylose exhaustion.

In relation to xylitol production it was observed a maximum peak from 48 h of fermentation to the experiments 17 and 18, presenting a conversion percentage of approximately 87 %, whereas for experiment 19 this peak occurred with 36 h, and an average 88.5 %, after those periods it was found a decline in the xylitol production.

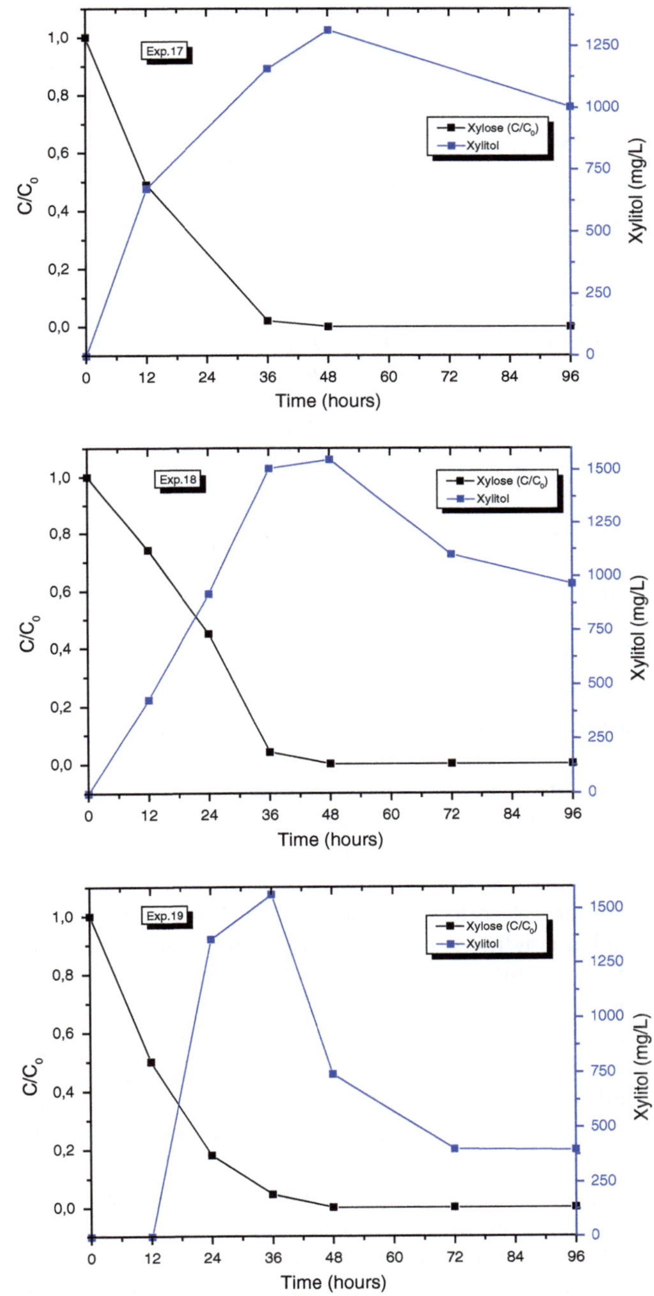

Fig. 8 Kinetic profiles of fermentations of the xylose consumption and xylitol production by using the yeast CCT 3544 *C. guilliermondii*, exp. 17–19

Table 5 2^4 planning matrix and experimental results obtained during the bioconversion of xylose to xylitol by CCT 3544 *Candida guilliermondii*

Experiments	Xylitol (mg/L)	$Y_{P/S}$ (%)	η (%)	Y (%)	Q_p (g/Lh)	$t_{max}X$ (h)
1	2754.66	90.709	98.920	99.916	0.108	12
2	2222.50	90.111	98.267	99.941	0.081	12
3	1433.37	86.823	94.682	100.00	0.028	36
4	743.90	80.025	87.268	98.235	0.018	24
5	2296.31	89.063	97.124	98.031	0.097	12
6	2953.19	88.882	96.927	99.708	0.118	12
7	1710.86	89.858	97.992	100.00	0.042	24
8	936.42	86.248	94.054	99.834	0.026	12/24
9	962.76	87.769	95.713	99.800	0.061	12
10	1114.96	85.047	92.744	99.050	0.082	12
11	1616.28	87.077	94.958	99.869	0.056	24
12	1436.13	86.018	93.804	97.984	0.045	12
13	976.99	89.323	97.408	99.965	0.047	12
14	772.93	89.012	97.069	100.00	0.022	24
15	1730.91	85.965	93.746	99.766	0.066	24
16	731.71	91.597	99.888	97.323	0.032	12
17	1313.99	91.292	99.556	100.00	0.056	12
18	1548.89	90.738	98.951	100.00	0.042	36
19	1566.90	89.775	97.900	99.977	0.057	24

4.5.2 Calculations of the Kinetic Parameters of Fermentation

Were calculated in this work, the kinetic parameters for all the experiments of the 2^4 experimental planning with three replications at the center point, and it was found the influence of the input variables on the kinetic parameters.

In Table 5 are found the values of the fermentative parameters obtained during the bioconversion of xylose to xylitol by CCT 3544 *C. guilliermondii*, using the different cultivation conditions corresponding to the planning in question, in the times of maximum xylitol production ($t_{max}X$).

It can be seen in Table 5, the maximum xylitol values obtained in each experiment and their respective kinetic parameters. The combination of independent variable levels (conditions of culture medium) provided some variations in the performance of the yeast CCT 3544 *C. guilliermondii*. It is found that the values of xylitol ranged in the 19 experiments with peak concentrations from 2953.19 to 731.71 mg/L, volumetric productivity from 0.118 to 0.032 g/L h, conversion factor of xylose–xylitol, 88.88–91.59 %, corresponding to a efficiency of conversion of 96.93–99.88 % were found in the experiments 6 and 16, respectively, in which were employed the highest levels of ammonium sulphate (3.0 g/L), and rice bran extract (20.0 g/L) in both experiments; lower levels of calcium chloride (0.5 g/L), and pH (4.0) for the experiment 6 and higher levels of calcium chlorides (2.0 g/L), and pH (6.0) for the experiment 16 at 12 h of fermentation.

Canilha et al. [57] evaluated the wheat straw for biotechnological production of xylitol and obtained maximum xylitol values (22.5 and 21.0 g/L), volumetric productivity (0.31 and 0.29 g/L h) and conversion factor in xylitol (0.45 and 0.50 g/g), efficiency of conversion of 49.07 and 54.53 %, for lower levels of rice bran (5.0 g/L) and the initial xylose concentration (60 g/L) and higher pH level (6.0) at 72 h of fermentation.

Cândido [58] found, in study of hydrolysate of the barley straw for xylitol production by *C. guilliermondii*, value of xylitol 49.68–18.15 g/L (trials 1 and 10) volumetric productivity 0.69–0.25 g/L h for lower levels of rice bran (5.0 g/L) and calcium chloride (0.0 g/L), lower level of ammonium sulphate (1.0 g/L) and initial xylose concentration (60.0 g/L) for trial 1 and the higher levels of ammonium sulphate (3 g/L) and initial xylose concentration (80 g/L) for trial 10, in 72 h of fermentation.

5 Conclusions

The *C. guilliermondii* was able to consume xylose in all fermentations carried out with greater xylitol production of 2953.19 mg/L in 12 h of fermentation.

The best culture conditions were 1 g/L of ammonium sulfate, 0.5 g/L of calcium chloride, 5 g/L of rice bran and pH of 4, obtaining the volumetric productivity of xylitol, factor, and efficiency of xylose conversion to xylitol of 0.108 g/L h, 90.70 and 98.99 %, respectively.

References

1. Leitão, V.F., Gottschalk, L.M.F., Ferrara, M.A., Nepomuceno, A.L., Molinari, H.B.C., Bom, E.P.S.: Biomass residues in Brazil: availability and potential uses. Waste Biomass Valorization **1**, 65–76 (2010)
2. Michel, A.C.S., Flôres, S.H., Hertz, P.F., Matos, G.S., Ayub, M.A.Z.: Production of ethanol from soybean hull hydrolysate by osmotolerant *Candida guilliermondii* NRRL Y-2075. Bioresour. Technol. **99**(8), 2898–2904 (2008)
3. Wei, J., Yuan, Q., Wang, T., Wang, L.: Purification and crystallization of xylitol from fermentation broth of corncob hydrolysates. Front. Chem. Eng. China **4**, 57–64 (2010)
4. Galbe, M., Zacchi, G.: Production of ethanol from lignocellulosic materials—bioethanol from sugarcane: R &D to productivity and sustainability. Luis Augusto Barbosa Cortez, São Paulo: Parte 4, Cap.12, Blucher, 697–716 (2010)
5. Oliveira, V.H.D., Andrade, A.P.S.: Integrated cashew production. Opening doors for quality. http://www.cnpat.embrapa.br/pif/artigo1/agroanalyse/index.html. Accessed on 07 April 2012 (In Portuguese)
6. Rodrigues, J.A.R.: Ingenuity to the biorefinery. The sugar mill as an industrial undertaking for the generation of biofuel and biochemical products. New Chem. **1**, 1–13 (2011). (In Portuguese)
7. Sheehan, J.J.: Biofuels and the conundrum of sustainability. Curr. Opin. Biotechnol. **20**, 318–324 (2009)

8. Canakci, M., Sanli, H.: Biodiesel production from various feed stocks and their effects on the fuel properties. J. Ind. Microbiol. Biotechnol. **35**, 431–441 (2008)
9. Kumar, R., Singh, S., Singh, O.V.: Bioconversion of lignocellulosic biomass: Biochemical and molecular perspectives. J. Ind. Microbiol. Biotechnol. **35**, 377–391 (2008)
10. Sun, Y., Cheng, J.: Hydrolysis of lignocellulosic materials for ethanol production: a review. Bioresour. Technol. **83**(1), 1–11 (2002)
11. Mussato, S.I., Teixeira, J.A.: Lignocellulose as raw material in fermentation processes. Appl. Microbiol. Microb. Biotechnol. 897–907 (2010)
12. Stambuk, B.U., Eleutherio, E.C.A., Florez-Pardo, L.M., Souto-Maior, A., Bom, E.P.S.: Brazilian potential for biomass ethanol: Challenge of using hexose and pentose co-fermenting yeast strains. J. Sci. Ind. Res. **67**, 918–926 (2008)
13. Morais, S.A.L., Nascimento, E.A., Melo, D.C.: Análise da madeira de *Pinus oocarpa.* Parte—Estudo dos constituintes macromoleculares e extrativos voláteis. Tree magazine **29**(3), 461–470 (2005). (In Portuguese)
14. Cruz, J.M., Domínguez, J.M., Domínguez, H., Parajó, J.C.: Preparation of fermentation media from agricultural wastes and their bioconversion to xylitol. Food Biotechnol.**14**(1–2), 79–97 (2000)
15. Santos, F.A., de Queiroz, J.H., Colodette, J.L., Fernandes, S.A., Guimarães, V.M., Rezende, S.T.: Potential straw cane sugar for ethanol production. New Chem. **35**(5), 1004–1010 (2012). (In Portuguese)
16. Shoudham, V.P., Rodriguez, D.R., Rocha, G.J.M., Taherzadeh, M.J., Martin, E.C.: Acetosolv delignification of marabou (*Dichrostachys cinerea*) wood with and without acid prehydrolysis. For. Stud. China **13**(1), 64–70 (2011)
17. Silva Neto, J.M., Silva, F.L.H., Lima, E.E., Torres Neto, A.B., Lima, F.C.S.: Analysis of acid pretreatment of the cashew apple bagasse. In: Brazilian Congress of Scientific Initiation of Chemical Engineering, 9, Maringá, Paraná, 2011, pp. 1–6, (In Portuguese)
18. Silva, D.D.V., Mancilha, I.M., Silva, S.S., Felipe, M.G.A.: Improvement of biotechnological xylitol production by glucose during cultive of *Candida guilliermondii* in sugarcane bagasse hydrolysate. Braz. Arch. Biol. Technol. **50**, 207 (2007)
19. Silva, J.P.A., Mussatto, S.I., Roberto, I.C.: The influence of initial xylose concentration, agitation and aeration on ethanol production by *Pichia stipitis* from rice straw hemicellulosic hydrolysate. Appl. Biochem. Biotechnol. **162**, 1306–1315 (2010)
20. Rocha, M.V.P., Rodrigues, T.H.S., Melo, V.M.M., Gonçalves, L.R.B., Macedo, G.R.: Cashew apple bagasse as a source of sugars for ethanol production by *Kluyveromyces marxianus* CE025. J. Ind. Microbiol. Biotechnol. **38**(8), 1099–1107 (2011)
21. Dos S. Lima, F.C., da Silva, F.L.H., Gomes, J.P., Silva Neto, J.M.: Chemical composition of the cashew apple bagasse and potential use for ethanol production. Adv. Chem. Eng. Sci. **2**, 519–523 (2012)
22. Aragão, R.F.: Drying slices of cashew (*Anacardium Occidentalel*), tray dryer. Thesis, Federal University Campina Grande, 126p (2007) (In Portuguese)
23. Agostini-Costa, T.S., Vieira, R.F., Naves, R.V.: Cashew, tropical identity that exude shealth. Brazilian Society of Forest Engineers (2006). www.cenargen.embrapa.br/cenargenda/divulgacao2006/sbef220106.pdf/. Accessed on July 2012 (In Portuguese)
24. Mazzetto, S.E., Lomonaco, D., Mele, G.: Oil of cashew nuts: opportunities and challenges in the development and industrial sustainability. New Chem. **32**(3), 732–721 (2009) (In Portuguese)
25. Lopes, M.M.A., Moura, C.F.H., Aragão, F.A.S., Cardoso, T.G., Eneas Filho, J.: Physical characterization of stems of dwarf cashew clones at different stages of maturation. Agron. Sci. J. **42**(4), 914–920 (2011) (In Portuguese)
26. Kumar, P., Barrett, D.M., Delwiche, M.J., Stroeve, P.: Methods for pretreatment of lignocellulosic biomass for efficient hydrolysis and biofuel production. Ind. Eng. Chem. Res. **48**(8), 3713–3729 (2009)
27. Gírio, F.M., Fonseca, C., Carvalheiro, F., Duarte, L.C., Marques, S., Bogel-Lukasik, R.: Hemicelluloses for fuel ethanol: a review. Bioresour. Technol. **101**(13), 4775–4800 (2010)

28. Talebnia, F., Karakashev, D., Angelidaki, I.: Production of bioethanol from wheat straw: an overview on pretreatment, hydrolysis and fermentation. Bioresour. Technol. **101**(13), 4744–4753 (2010)
29. Silverstein, R.A., Chen, Y., Sharma-Shivappa, R.R., Boyette, M.D., Osborne, J.: A comparison of chemical pretreatment methods for improving sacarification of cotton stalks. Bioresour. Technol. **98**(16), 3000–3011 (2007)
30. Mussato, S.I., Roberto, I.C.: Optimal experimental condition for hemicellulosic hydrolyzate treatment with activated charcoal for xylitol production. Biotechnol. Prog. **20**(1), 134–139 (2004)
31. Saha, B.C., Cotta, M.A.: Ethanol production from alkaline peroxide pretreated enzymatically saccharified, wheat straw. Biotechnol. Prog. **22**, 449–453 (2006)
32. Parazzi, C.: Raw materials of fermentation ethanol (2006). http://www.cca.ufscar.br/~vico/2%20monitoramento/2%20Materia%20prima.pdf. Accessed on 28 Nov 2012 (In Portuguese)
33. Misra, S., Gupta, P., Raghuwanshi, S., Dutt, K., Saxena, R.K.: Comparative study on different strategies involved for xylitol purification from culture media fermented by *Candida tropicalis*. Sep. Purif. Technol. **78**(3), 266–273 (2011)
34. Wei, J., Yuan, Q., Wang, T., Wang, L.: Purification and crystallization of xylitol from fermentation broth of corncob hydrolysates. Front. Chem. Eng. China **4**, 57–64 (2010)
35. Mäkinen, K.K.: Can the pentitolhexitol theory explain the clinical observations made with xylitol. Med. Hypotheses **54**(4), 603–613 (2000)
36. Hyvönen, L., Törmä, R.: Examination of sugars, sugar alcohols, and artificial sweeteners as substitutes for sucrose in strawberry jam. Keeping quality tests. J. Food Sci. **48**(1), 186–192 (1983)
37. Vernacchio, L., Vezina, R.M., Mitchell, A.A.: Tolerability of oral xylitol solution in young children: implications for otitis media prophylaxis. Int. J. Pediatr. Otorhinolaryngol. **71**, 89–94 (2007)
38. Pereira, R.S., Mussatto, S.I., Roberto, I.C.: Inhibitory action of toxic compounds present in lignocellulosic hydrolysates on xylose to xylitol bioconversion by *Candida guilliermondii*. J. Microbiol. Biotechnol. **38**, 71–78 (2011)
39. Mussatto, S.I., Roberto, I.C.: Xilitol: Edulcorante com efeitos benéficos para a saúde humana. Braz. J. Pharm. Sci. **38**(4), 401–413 (2002)
40. Cassales, A.R., Souza-Cruz, P., Ayub, M.A.: Comparison of the ethanol production from hydrolysed soy hull sand synthetic medium used using microbial consortia. In: Symposium on Applied Microbiology, 3, Porto Alegre. Porto Alegre, 2009. CD-Rom (In Portuguese)
41. Cunha, M.A.A., Silva, S.S., Carvalho, W., Santos, J.C.: Use of immobilized cells of PVA gel: a new strategy for the biotechnological production of xylitol logic from bagasse of sugar. Seminary **26**, 59–68 (2005). (In Portuguese)
42. Sarrouh, B.F., Silva, S.S.: Evaluation of the performance of a three-phase fluidized bed reactor with immobilized yeast cells for the biotechnological production of xylitol. Int. J. Chem. Reactor Eng. **6**, 1–15 (2008)
43. Faveri, D., Torre, P., Perego, P., Converti, A.: Optimization of xylitol recovery by crystallization from synthetic solutions using response surface methodology. J. Food Eng. **61**(3), 407–412 (2004)
44. Winkelhausen, E., Kuzmanova, S.: Microbial conversion of D-xylose to xylitol. J. Ferment. Bioeng. **86**(1), 1–14 (1998)
45. de Castro, A.M., Pereira Júnior, N.: Production, properties and application of cellulases in the hydrolysis of organic residues. New Chem. **33**, 181–188 (2010). (In Portuguese)
46. Pinheiro, T.L.F., Menoncin, S., Domingues, N., Oliveira, D., Treichel, H., Luccio, M., Freire, D.M.G.: Production and partial characterization of lipase from *Penicillium verrucosum* obtained by submerged fermentation of conventional and industrial media. Sci. Food Technol. **28**(2), 444–450 (2008)
47. Hahn-Hägerdal, B., Galbe, M., Gorwa-Grauslund, M.F., Lidén, G., Zacchi, G.: Bio-ethanol—the fuel of tomorrow from the residues of today. Trends. Biotechnol. **24**(12), 549–556 (2006)

48. Prakasham, R.S., Sreenivas, R.R., Hobbs, P.J.: Current trends in biotechnological production of xylitol and future prospects. Curr. Trends Biotechnol. Pharm. **3**(1), 8–36 (2009)
49. Viegas, M.C.: Optimization of fermentation system using continuous reactors tower type and yeast characteristics with flocculants. UNICAMP, Campinas, Thesis, 150p (2003) (In Portuguese)
50. Gee, D.A.: Modelling, optimal control, state estimation, and parameter identification applied to a batch fermentation process. UC, Colorado. Thesis (1990) (In Portuguese)
51. Luong, J.H.T.: Kinetics of ethanol inhibition in alcohol fermentation. Biotechnology and Bioengineering, 280–285 (1985)
52. de Lima, E.E., da Silva, F.L.H., dos S. Lima, F.C.: Determination of the efficiency of the solvents in the extractive content of cashew apple pomace. In: National Symposium on Bioprocesss, 18, SINAFERM 2011, Caxias do Sul, CD-Rom (In Portuguese)
53. Nunes, B.R.P., Conrado, L.S.: Remoção de furfural e HMF utilizando argila vermiculita como Adsorvente In: National Symposium on Bioprocesss—SINAFERM, 18, 2011, Caxias do Sul, UCS, 2011, pp. 1–6. CD-Rom (In Portuguese)
54. Fonseca, C.R., Faria, L.F.F.: Production of ethanol and xylitol from hemicellulosic hydrolyzate detoxified. In: Brazilian Congress of Chemical Engineering, 19, Búzios/ RJ, COBEQ, 2012, pp. 4820–4829 (In Portuguese)
55. Bier, M.C.J., Maranhão, L.T., Azevedo, J.A.M., Silva Junior, L.S.: Growth and consumption of xylose by *Candida guilliermondii* in submerged fermentation of xylose using bagasse cane sugar. Evid. Mag. **7**(2), 119–130 (2007)
56. Moutta, R.O., Rocha, G.J.M., Silva, S.S.: Optimization of the sugar cane straw hydrolysis conditions aiming the attainment of bioethanol. In: European Congress on Biotechnology, 14, Barcelona. New Biotechnology (2009)
57. Canilha, L., Carvalho, W., Silva, J.B.A.: Xylitol bioproduction from wheat straw: Hemicellulose hydrolysis and hydrolizate fermentation. J. Sci. Food Agric. **86**, 1371–1376 (2006). (In Portuguese)
58. Cândido, E.J.: Economic feasibility study of xylitol production from hemicellulose hydrolyzate of barley straw. USP, Lorena, Thesis, 157p (2008) (In Portuguese)
59. Bar, A.: Xylitol. In: O'Brein Nabors, L., Gelardi, R.C. (eds.) Alternative Sweeteners, 2nd edn., pp. 349–379. Marcel Dekkor Inc., New York (1991)

MIX
Papier aus verantwortungsvollen Quellen
Paper from responsible sources
FSC® C105338

If you have any concerns about our products,
you can contact us on
ProductSafety@springernature.com

In case Publisher is established outside the EU,
the EU authorized representative is:
Springer Nature Customer Service Center GmbH
Europaplatz 3, 69115 Heidelberg, Germany

Printed by Libri Plureos GmbH
in Hamburg, Germany